"Robot Woman from Outer Space"

P9-CJM-417

Computer software is
a kind of automaton —
bodiless...

Finney's 5 Against
the House used a
Turk-like device!
— A machine pretends
to be a man.

Imbuing life — volition —
into objects such as
weapons (swords), chalices,
and the like.

Finished: 12/15/2007

EDISON'S EVE

EDISON'S EVE

A Magical History of the
Quest for Mechanical Life

GABY WOOD

ALFRED A. KNOPF

NEW YORK

2002

THIS IS A BORZOI BOOK
PUBLISHED BY ALFRED A. KNOPF

Copyright © 2002 by Gaby Wood

www.aaknopf.com

Originally published in Great Britain as *Living Dolls*
by Faber and Faber Limited, London.

Knopf, Borzoi Books, and the colophon are
registered trademarks of Random House, Inc.

Library of Congress Cataloging-in-Publication Data
Wood, Gaby.
Edison's Eve : a magical history of the quest for mechanical life / Gaby Wood.
p.   cm.
ISBN 0-679-45112-9 (alk. paper)
1. Robots—Design and construction—History.   2. Artificial intelligence.
I. Title.
TJ211.W65   2002
629.8'92—dc21   2002025467

Manufactured in the United States of America
First American Edition

For my parents

ART met her sister NATURE late,
And seeing her at ease,
Invited her to take a seat
At her Androides;

Dame NATURE went—was pleas'd at first,
And warmly praised her sister;
Then laughing, till she nearly burst,
In seeming rapture kiss'd her.

But as the wond'rous figures work'd
She look'd a little serious,
Whilst envy in her bosom lurk'd—
Her brow became imperious.

"How's this!" to ART she loudly said,
"How's this! ungrateful creature!
Profanely thou hast dar'd to tread
Thus in the walks of NATURE.

"I prithee, base, usurping wench,
No more these freedoms take;
If thus my province thou intrench—
Thou'lt men and women make."

Anonymous poem published in the *Bath Herald*, Saturday,
28 January 1797, on the occasion of "Mr. Haddock's Exhibition
of Androides."

Contents

List of Illustrations   *xi*
Introduction   *xiii*

ONE
The Blood of an Android   *3*

TWO
An Unreasonable Game   *60*

THREE
Journey to the Perfect Woman   *111*

FOUR
Magical Mysteries, Mechanical Dreams   *164*

FIVE
The Doll Family   *213*

Epilogue   *264*

Acknowledgements   *267*
Bibliography   *271*
Index   *291*

Illustrations

1. Vaucanson's automata, as exhibited in 1739. *Reprinted with permission of the British Library.*
2. The purported guts of Vaucanson's duck.
3 and 4. Two photographs, taken in the 1890s, are thought to represent the skeleton of Vaucanson's duck, and the large pedestal in which its mechanism was hidden. © *Musée des arts et métiers-Cnam, Paris / Photograph P. Fallgot / Seventh Square.*
5, 6, and 7. The front and back views of Kempelen's chess player, published to show that no man could possibly be hidden inside the machine, and an attempt to expose Kempelen, drawn by Joseph Friedrich Freiherr zu Racknitz. *Princeton University Library, Cook Chess Collection, Department of Rare Books and Special Collections.*
8. A talking doll made in 1890 by Thomas Edison. *U.S. Department of the Interior, National Park Service, Edison National Historic Site.*
9. A label from the packaging of a talking doll: the "talking number" indicates which rhyme she could recite. *Reprinted with permission of the British Library.*
10. An engraving of Edison's talking doll, as printed on the cover of *Scientific American* in April 1890. *U.S. Department of the Interior, National Park Service, Edison National Historic Site.*
11. The manufacture of talking dolls at the Edison factory in

New Jersey. *U.S. Department of the Interior, National Park Service, Edison National Historic Site.*

12. The magician John Nevil Maskelyne, pictured with his automaton, Psycho. *Reprinted with permission of the British Library.*

13. A still from Georges Méliès's 1902 trick film *The Man with the Rubber Head. Courtesy of the Méliès estate.*

14. Georges Méliès, living in oblivion towards the end of his life. *Courtesy of the Méliès estate.*

15. The Doll Family in the 1930s with the silent movie star Harold Lloyd. *Courtesy of Tiny Doll.*

16. The Doll Family in the 1920s, dressed in the glamorous outfits they wore for their performances with the Ringling Brothers Circus. *Circus World Museum, Baraboo, Wisconsin.*

17. Harry Doll with Jack "Sky High" Earle. *Circus World Museum, Baraboo, Wisconsin.*

18 and 19. Front and back views of Jacquet-Droz's eighteenth-century writing automaton. *Photographs by Anne de Tribolet, courtesy of the Musée d'art et d'histoire, Neuchâtel.*

20. Kismet, as seen at the humanoid robotics lab at the Massachusetts Institute of Technology. *Peter Menzel / Science Photo Library.*

21. Hadaly, as seen at the Takanishi lab at Waseda University in Tokyo. *Shigeki Sugano Lab, Humanoid Robotics Institute, Waseda University.*

Introduction

Once a month, in a lakeside town in Switzerland, two little boys perform feats of great dexterity. These prodigies, who look no older than toddlers, are dressed for the occasion in identical velvet jackets and silk pantaloons. Their faces are doll-like and blank; their bare feet dangle some way off the ground. The first boy begins by dipping his quill pen in a tiny ink well at the side of his desk. He shakes it twice, then methodically moves his hand across the paper and starts to trace the letters in his message. Meanwhile, his twin works on a sketch. He slowly draws a head in profile, then drops his chin and blows away the dust from his pencil. While the boys perform their dutiful activities before a small but avid crowd, they are turned to face the wall: their clothes are pulled away and their spines prized open. Inside each child is a moving piece of golden clockwork.

Introduction

These children have not aged for over 200 years. The draughtsman still draws portraits of Louis XV and George III; the writer still communicates to its audience an eerie philosophical joke: *"I think,"* it writes, *"therefore I am."* They were first exhibited here, in Neuchâtel, by their inventors, Pierre Jaquet-Droz and his son Henri-Louis, in 1774. It was said that people came to see them as if on a pilgrimage, from miles away, and ever since then these artificial beings have enchanted, frightened, and perplexed their viewers.

In 1776, another Jaquet-Droz android, a "Musical Lady" that played the harpsichord, was exhibited in London. As she played the five tunes in her repertoire, her eyes would move coyly from side to side, and her bosom would heave lightly, as if she were breathing. The machine was advertised on the poster as "a vestal virgin with a heart of steel," but one member of the audience thought her heart might be otherwise. A journalist who went to the exhibition reported that "she is apparently agitated with an anxiety and diffidence not always felt in real life." The description seems far-fetched, but perhaps it was merely misplaced. Clearly, there was an anxiety present in the situation—an anxiety that all androids, from the earliest moving doll to the most sophisticated robots, conjure up. Mixed in with the magic and the marvel is a fear: that we can be replicated all too easily, and that we are uncertain now of what it is that makes us human. In other words, in saying that the android was anxious, the reporter had projected onto the jerkily breathing machine the emotions it elicited in him.

His reaction is a perfect example of what Sigmund Freud called "the Uncanny," the feeling that arises when there is an "intellectual uncertainty" about the borderline between

A statue becomes alive
A doll or an image in a photo or movie comes alive
A puppet, a mechanical device – machines themselves (Christine)
plants, ...

*

the lifeless and the living. It is triggered in particular, Freud wrote, by "waxwork figures, ingeniously constructed dolls and automata." A child's desire for a doll to come to life may become, in adulthood, a fear.

The eighteenth-century journalist was not the only one who was worried. When Pierre Jaquet-Droz exhibited his writing automaton in Spain, he was accused of heresy; both the man and the machine were imprisoned for a time by the Spanish Inquisition. Decades later, Mary Shelley passed through Neuchâtel on her six-week-long tour of Europe. It is thought that she may have seen Jaquet-Droz's androids then, two years before she wrote *Frankenstein*, and it seems fitting that the writer should have been here, watching her inanimate counterpart at work, and dreaming up a monstrous fiction about artificial life.

For centuries, there have been men who might have claimed Shelley's subtitle—"the modern Prometheus"—for themselves, and others who were given the label, either in admiration or as a warning against tempting God. A number of these manufacturers of artificial life were documented in *Lives of the Necromancers*, the last book written by Mary Shelley's father, William Godwin. And some of them were said in Shelley's novel to have influenced Victor Frankenstein in his youth.

Godwin told the story of Daedalus, the master craftsman of classical mythology, who was said to have made dolls with moveable limbs. Aeschylus wrote about his living statues, and Socrates said of these sculptures that "if they were left untethered, they might take off, giving you the slip like a runaway slave." Archytas of Tarentum, a contemporary of Plato credited with the invention of the crane, is supposed to have constructed a flying pigeon out of wood. Around 150 B.C., Hero of Alexandria (the man who invented the syringe) came up with his own death-defying automaton: it was a simulated human being whose head (protected by an ingenious system of cogs and wheels) could not be severed from its neck, no matter how many times you passed a knife through it. Albertus Magnus, a thirteenth-century Dominican monk, spent thirty years building an artificial man out of brass. He gave the android the power of speech, and made it his servant, but its speech was to be its downfall. Albertus's student Thomas Aquinas became so furious at the brass man's chatter that he beat it to pieces with a hammer.

Not all attempts to create artificial life were mechanical. During the Renaissance, a number of astrologers and alchemists developed recipes for more magical creations. Cornelius Agrippa believed, along with many of his fellow sorcerers, that humans could be grown from mandrake roots. His contemporary Paracelsus published instructions for the manufacture of a "homunculus," or miniature man. Human semen, Paracelsus suggested, should be put into an airtight jar and buried in horse manure for forty days. After this, it was to be "magnetized," then preserved at the temperature of a mare's womb and fed human blood for forty weeks. A small, fully formed person was thought to emerge after this procedure. (In Goethe's *Faust,* the doctor attempts this very trick.)

Despite the ancient history of artificial life, I have chosen to begin this story of androids and their makers at a later, critical moment: when the ambitions of the necromancers were revived in the well-respected name of science. In the eighteenth century, an interest in anatomy, advances in the design of scientific instruments, and a fondness for magic tricks meant that automata were thought of as glorious feats of engineering, or philosophical toys. As Umberto Eco has put it, the inventors of that period "substituted mechanics for the forces of evil." And yet, the shift was not as straightforward as it seemed: the rational scientists who constructed these celebrated objects often harboured ambitions beyond the bounds of reason. One would name his clockwork figure after a daughter who had died, as if there were a kind of resurrection involved; another used his spare time to make artificial limbs; yet another covered the hands of his flute-playing automaton in skin, and spent his life trying to make a machine that could bleed. Each of these projects blurred the line between man and machine, between the animate and the inanimate. The madness left over from darker times was all the more disturbing for being hidden beneath the mask of enlightenment.

Indeed, the dangers of animation were, if anything, accentuated by the technical virtuosity the Enlightenment so admired. The sophisticated mechanisms of eighteenth-century automata contained within them a unique philosophical difficulty: by its very nature, clockwork is the antithesis of our mortal selves ("our mechanisms defy time" is one of the phrases the Jaquet-Droz automaton writes). Time is wrapped up in the android in a way that is directly opposed to the way in which we are wrapped up in time. Man is subject to time, to its inevitable march towards death, whereas the clockwork automaton merely marks time without falling prey to it.

John Cohen, a psychology professor and the author of a history of robots, has suggested that although a robot may fall out of order, it cannot bring about its own destruction. A robot can never commit suicide, he writes, because "true suicide implies a foreknowledge of death and some idea of its significance, and this is a privilege of man." In Karel Capek's 1921 play *Rossum's Universal Robots,* from which the word "robot" originally derives, a woman asks a female robot if she is afraid of dying. The robot does not simply say no. She says, in a perfect illustration of Cohen's point, "I cannot tell." Death means nothing to her.

But although androids have no understanding of death, they are themselves embodiments of it. Every time an inventor tries to simulate life mechanically, he is in fact accentuating his own mortality. He holds his creation in his hands, and finds, where he expected life, only the lifeless; the closer he comes to attaining his goal, the more impossible it reveals itself to be. Rather than being copies of people, androids are more like mementi mori, reminders that, unlike us, they are forever unliving, and yet never dead. They throw the human condition into horrible relief. Something of this is implied in the name given to the futuristic androids in *Blade Runner.* A "replicant," as they are called, is a combination of a replica and a revenant—as if they had returned from the grave in order to mimic humanity; as if death were inherent in the simulacrum.

In the eighteenth century, if you looked up "automaton" in Diderot's *Encyclopédie,* you would have found a description of Jacques de Vaucanson's artificial duck, made in 1739. Now, the *Oxford English Dictionary* offers a reflection of the automaton's subsequent history. Aside from the basic definition—a machine that contains within itself the power of motion—there are contradictory possibilities for what an automaton can mean. It is either "a figure which simulates the action of a living being," or, conversely, "a human being acting mechanically in a monotonous routine." At the turn of the eighteenth to the nineteenth century, makers of automata turned away from constructing curiosities in order to design the machines that would replace human labour. From that point on, men were required to operate, repetitively, "mechanically," the objects that had usurped them. When Karel Capek coined the word "robot" from the Czech for "to work," he was showing how the line between man and machine was blurred once again: instead of machines being made to look like men, men became more like machines.

It was at this moment, the scholar Andreas Huyssen has argued, that the culture of the android moved into fiction. In E. T. A. Hoffmann's story "The Sandman," and Villiers de l'Isle-Adam's novel *The Eve of the Future,* men fall in love with androids, believing them to be perfect women. In both cases their mistake is fatal. And so, in the shift from the real to the imaginary and from the playful to the destructive, androids ceased to be male and became, more often than not, female.

It should come as no surprise that the inventors in this book are men (prone perhaps to what some psychologists have called "womb envy"). They are latter-day Pygmalions, whose statues come to life and become not Galatea but Pandora, the destructive "manufactured maiden" of Greek mythology, sent to earth by Zeus as punishment for Prometheus's transgression.

If they were rarely meant to be women, however, the automata of the Enlightenment were frequently designed in the image of children. Some inventors intended their objects to be artificial forms of an eighteenth-century ideal—the child as a blank slate, the purest being. The Jaquet-Droz figures conduct their marvellous activities barefoot, illustrating a belief, held by their contemporary Jean-Jacques Rousseau, that children would learn more freely if unhampered by shoes. (They are living extensions of the young people in Chardin's paintings—naïve yet enlightened, peacefully engaged in study.) Other mechanical children, however, were prototypes: their creators wanted them to look young so that the mistakes resulting from their early efforts would be forgiven. In those cases the status of a child was given, not as a sign of perfection, but as an insurance against failure.

Exactly what, these child-automata may lead us to wonder, is wanted of a machine? Is it supposed to be as close as possible to a human being, or to improve on that, and become superhuman? In the quest for mechanical perfection, does perfection mean infallibility (as in the computer), or innocence (as in the child)? Curiously, contemporary research into artificial intelligence has returned to these issues. While once it was thought that true artificial intelligence entailed an infinite number of calculations (the computer as all-knowing adult), now more attention has been turned towards "learning machines," robots that start out as infants and pick things up as they go along. The tenets of Rousseau have been resurrected in the technological laboratory.

The first artificial intelligence laboratory was founded in the 1950s at the Massachusetts Institute of Technology. That lab has since developed new technology for surgical operations, it has wired the White House, and research conducted by the lab's director, Rodney Brooks, was used to build a planetary rover for the 1997 Pathfinder mission to Mars. In their humanoid robotics department, researchers are in the process of building some of the most sophisticated robots in the world.

At first sight, the robotics lab seems a straightforward enough place: it is clean and uncrowded, with a few people working at computers, and tall stacks of hard-drives furnishing each room. Robots are being repaired and improved everywhere—a mechanical arm lies on a table, someone is putting disembodied legs together. But when the robots are working, there is a creeping sense of a living presence. "Cog," a monumental metal torso on a three-foot pedestal, follows me around the room with its eyes—four camera lenses to simulate direct and peripheral vision, which are linked up to computer screens above its head. Cog's head moves up and down, and turns around on its neck. Its eyes swivel in their sockets, and hold me in a steady gaze. From the corner of my eye, as I look at Cog, I can see myself replicated four times on the screens behind it. Underneath those are myriad other screens, full of changing computer codes in digital green lettering.

"Kismet," a robot head that can express human emotions, is made to look more friendly. Its frame is much like Cog's, but it has large blue eyeballs, eyelids made of silver foil, bits of carpet for brows, twists of pink paper for ears, and red rubber tubing to make a mouth. Kismet is a cartoonish, mechanical gremlin. Each of its facial features moves according to Kismet's mood, depending on whether it is happy or sad or calm or angry. From time to time Kismet will make a little gurgling noise, like a child.

What these robots can do is extraordinary: Cog can recognize what a human face is, and has been taught to tell the difference between an animate and an inanimate object—exactly the distinction androids blur for their human viewers (how, I wonder, would it view itself?). Kismet, meanwhile, is the only robot in the world that has needs. When it wants to play, it is appeased when Cynthia Breazeal, its designer and "human caregiver," shows it a brightly coloured toy. When it yearns for human contact, it looks angry or sad until someone walks into the room. The idea is that both machines will learn by interacting with human beings: they were built to be "sociable," and Rodney Brooks has observed that people do interact with these robots much more than they do with computer simulations, even though a two-dimensional virtual reality might be visually more lifelike.

And yet, while an engineer or a computer scientist might be most interested in what they can do, what the robots cannot do tends towards the philosophical. For example, they have no sense of time. They have various kinds of memory—for data, and something similar to muscle memory, or reflexes—but they still do not have what Breazeal calls "autobiographical memory." They can recognize a face, but they cannot tell that it is the same face they recognized yesterday. As Brooks puts it, they "live in this weird world—there is no past, everything is the present."

The name given to the robots' mechanical joints lends them an unexpected anthropomorphism: they are called "degrees of freedom." The term simply describes the movements they are capable of (Cog, for example, has 22 degrees of freedom, in its arms, torso, neck, and eyes). But it is reminiscent of Karel Capek's *Rossum's Universal Robots*. In that story, robots form an army and destroy the human race. Here, there is an implied emancipation in the phrase "degrees of freedom," as if there were a kind of boiling point—a robot has only to reach a certain degree of freedom, and then it will be truly free, a master of human slaves.

The robots here are not designed to replace human beings, as Capek feared. Cog was named for humanoid cognition, rather than for its place in a larger machine. It was not made to perform a specific labour task, but is instead what Breazeal refers to as a "thought experiment," a way of "arriving at a better understanding of ourselves." Not unlike the eighteenth-century philosophical toy, the robots at MIT are concrete puzzles, dreamed up in order to answer questions about human beings. It is important, Breazeal says, that they are not human: they are more usefully seen as metaphors. "If anything," she explains, "you come away from a project like this full of awe, because humans are so much more complicated."

Yet it is Breazeal's robot, the emotion-machine, which might be seen to threaten us most. Now that computers can have an intelligence of sorts—they can play chess, for example—feelings are the last bastion of humanity. I suggest to Rod Brooks that Kismet is expressing emotions but not actually experiencing them. "But how do we know if it's experiencing them?" he counters. "It's a very emotion-laden question, because it gets at the specialness that we have. So it's very hard to give scientific answers."

This difficulty is in keeping with the many scientific-seeming experiments that have in fact been fraught with anxiety. The Turing test, put forward in 1950 by the mathematician and wartime cryptanalyst Alan Turing, provided the foundations for modern-day artificial intelligence. It consisted of an "imitation game," as Turing called it, in which the players were a human being and a computer, placed in separate rooms, both invisible to an examiner. The examiner would ask questions, the players' responses to which would come up in writing on a screen. The object of the game, for both players, was to prove to the examiner that they were human. If a computer succeeded in being indistinguishable in its responses from a human being, it would pass the test. Turing anticipated that computers would have reached a 30 per cent success rate by the year 2000. So far no computer has ever passed.

To distinguish a man or a woman from a machine has been the aim, or the fear, of all those who have observed androids for centuries. In that sense, the test is similar to the pamphlets written to expose the first attempt at artificial intelligence, the eighteenth-century Automaton Chess Player; and it is clearly the same as the "empathy test" in *Blade Runner,* designed to sift out the "replicants" from the humans. The anxiety underlying each of these situations is: why would we need a test to tell the difference between ourselves and robots? Are we really so similar?

When thinking about the Turing test, attention is usually focused on the computer, since the aim, for computer scientists, is to make their system "pass" for humans. But the other half of the test involves a human being competing against the computer, trying to answer questions in a way that proves he or she is alive. It is not the same as Kasparov competing against Deep Blue—that was about who won at chess; we knew to begin with that Deep Blue was a computer, and Kasparov a man. The Turing test, however, is a test of identity, and a test of life. The examiner must decide which of the two players is the human being and which is the machine. It's possible that the examiner might surrender, and decide both players must be human, but technically, if the game is won by the computer, the human being has lost. If the computer is declared human, then the human is rendered inanimate, disproven, dead.

Some of the research done at the AI lab at MIT is used in commercial ventures. In collaboration with the toy company Hasbro, Rod Brooks's company iRobot has manufactured "My Real Baby Doll," of which they sold 150,000 at Christmas. The doll, an ordinary-looking thing with a rubber face and limbs and a padded body filled with circuitry, costs $99. Like Kismet, it has certain needs: it gurgles when it wants its bottle, burps when it needs to be held over one shoulder, says "uh-oh" when a nappy change is in order.

Brooks demonstrates all this, cradling the artificial child, burping it, feeding it. The nappy soiling, Brooks shows me as he removes the doll's clothes, is virtual: there are light sensors implanted around the doll's crotch so that as soon as a change has been simulated, it will stop crying. Brooks originally wanted to design the doll so that it cried more and more hysterically until it got what it wanted, but Hasbro said this was unworkable. The doll had to remain at the level of entertainment, they insisted, rather than become an actual simulation of human life. Now the doll snores and wheezes for a while if you lay it down, then falls to sleep. It wakes up again automatically if you pick it up within half an hour; after that, you have to physically switch it back on.

Brooks explains that doll manufacture is "a very cutthroat

business." He once took three of these dolls to an international toy fair and one of them was stolen. After they were put on the market, someone from iRobot went to visit a toy company in Japan, and found "My Real Baby" taken to pieces on a sort of operating table, all eyeballs, dismembered limbs, and electronic chips.

Although what goes on at MIT is undeniably sophisticated and cutting-edge, the most extraordinary thing about it is how consistent it is with concerns that have been around for centuries. What we think of as the future—the subject of science fiction or the aim of the laboratory—is in fact very much rooted in the past. Brooks's doll is a throwback to Thomas Edison's nineteenth-century toy; Brooks speaks about uploading DNA, the melding of flesh and machines, which was the life's work of Vaucanson and eighteenth-century anatomists. The notion of artificial intelligence began with Wolfgang von Kempelen's chess automaton in 1769. The films that keep rehearsing these paranoias—*2001*, *Blade Runner*, *Westworld*, *Robocop*, *Terminator*, *AI*—were themselves born from automata, when the technology of nineteenth-century androids led Georges Méliès to invent trick photography in cinema, the first virtual reality.

When I ask Brooks about this continuity, he says that, even now, "all these questions"—questions about memory and consciousness and emotions; about what makes us human—"are fuzzy." "And maybe," he goes on, "it's that we don't yet know how to state the questions, in the same way that questions about the cosmos didn't make much sense before astronomy. So we can play around with these questions, but they don't make sense. When you push on the questions they all break, in some funny way. And maybe we're just too . . . ignorant. We're sitting here on this flat earth, contemplating the heavens above—if we think we're on a flat earth, we're just not asking the right questions."

History is full of these puzzles, about how far we seem to have come, and how little we have travelled. In each of the chapters in this book I have tried to revive such a puzzle—so that each story is a story about anxiety, and each one is also an idea. *Edison's Eve* deals with what troubles us when we are faced with certain versions of ourselves—bionic men, speaking robots, intelligent machines, or even just a doll that moves. The modern world is so full of artificial creatures that we dare not stop to think what it means to want to make a perfect copy of a human being. But behind each of these inventions is a single notion: that life can be simulated by art or science or magic. And embodied in each invention is a riddle, a fundamental challenge to our perception of what makes us human.

The territory is long familiar from the realms of fairy tale and science fiction, yet it is outside the reach of any reference book. A fiction can imagine a world of simulated souls and project onto it a set of terrifying possibilities. An encyclopaedia will tell you which copy was created when. But somewhere in between these accounts lies a true history, played out over centuries and made up of facts as well as fears. It is the story of the men who wanted to play God, and of their awe-struck public, who at the sight of every artificial life worried about the authenticity of their own. *Edison's Eve*, in other words, is the prehistory of a modern idea.

Even to love humbugs—art that is a stand in for reality

Introduction

When I say goodbye to Cynthia Breazeal, she tells me I should go to Japan, where robots are made to look more lifelike, and people are less worried about them usurping human faculties. At Waseda University in Tokyo, she says, there is a robot that is truly amazing: it plays the flute.

She doesn't seem to realize that a French inventor built just such a machine, over 250 years ago . . .

EDISON'S EVE

```
CHAPTER ONE
```

The Blood of an Android

To examine the causes of life, we must first have recourse to death.

—Mary Shelley, *Frankenstein*

He was sure it was to be his last trip. The philosopher René Descartes had been summoned by Queen Christina of Sweden, who wanted to know his views on love, hatred, and the passions of the soul; but although he was happy to correspond with the Queen, Descartes was loath to become part of her court. He felt, he said, that "thoughts as well as waters" would freeze over in Sweden and, since that winter was particularly harsh, he believed he would not survive the season. He even feared, he wrote to a friend, "a shipwreck which will cost me my life." But Christina's whim was his command. Filled with foreboding, he packed his bags, taking all of his manuscripts with him.

He was travelling, he told his companions, with his young daughter Francine; but the sailors had never seen her, and, thinking this strange, they decided to seek her out one day, in the midst of a terrible storm. Everything was out of place; they could find neither the philosopher nor the girl. Overcome with curiosity, they crept into Descartes's quarters. There was no one there, but on leaving the room, they stopped in front of a mysterious box. As soon as they had opened it, they jumped back in horror: inside the box was a doll—a living doll, they thought, which moved and behaved exactly like a human being. Descartes, it transpired, had constructed the android himself, out of pieces of metal and clockwork. It was indeed his progeny, but not the kind the sailors had imagined: Francine was a machine. When the ship's captain was shown the moving marvel, he was convinced, in his shock, that it was some instrument of dark magic, responsible for the weather that had hampered their journey. On the captain's orders, Descartes's "daughter" was thrown overboard.

It's hard to know if this story is true. Descartes did go to Sweden, and did, as he had feared, die there, six months later. He had, in fact, attempted to build some automata earlier in his life (one of his correspondents reported that Descartes had plans for "a dancing man, a flying pigeon, and a spaniel that chased a pheasant"), and he continued to be interested in mechanical toys. But the events on the ship read like a too-perfect fable—about science falling prey to the God-fearing crowd, about the threatening, uncanny power of machines, about the rational philosopher who has an almost superstitious relation to the product of his own mind: he names it, he calls it his daughter—and whether or not the story is made up of literal facts, it must, in a sense, be true to some metaphorical purpose: what is the use of telling it? (It has been told many times since Descartes's death.)

Descartes did have a daughter, and her name was Francine, but by the time this story is said to have taken place, Francine had been dead for many years. She was born in 1635, to a servant named Hélène Jans, whom Descartes never married. She lived with her father, at least some of the time, in the Netherlands, and he was planning to take her with him to France, when she died of scarlet fever, at the age of five. He told a friend that her death was the greatest sorrow of his life.

Seen from this angle, the Descartes of the story comes across, not as the reasoning philosopher, but as a fallible human being, distraught, nine years later, by the death of his child. Unable to mourn her, he constructs a simulacrum of the girl, gives it the power of motion, names it after her. If death was, as the following century liked to call it, "suspended animation," then Descartes, in animating this doll, had defied mortality and resurrected his daughter. Perhaps he had even done something, symbolically, for his own lifespan. Some years earlier, when he had been focusing his work on medicine, Descartes had written that he thought he could live to be a hundred. Francine died shortly after that. The making of the doll might be seen as an attempt to counter the terrible dashing of his hopes of extended life; and it seems fitting then that the ageless clockwork figure should have been destroyed on the trip where he was eventually to meet his end. This would suggest that the sailors might have been right to fear the object, not in itself, but because of Descartes's strange attachment to it.

Perhaps, however (since we cannot be sure of Descartes's intentions), the story can only be understood as one put about by later generations, in which case what is interesting is the confusion of the culture behind it. The fable is a new configuration, built up out of anxiety. It describes, in the mind of the storyteller, and in that of the audience, an uncertainty about categories. What is the difference between a person and a machine? Where is the line between a child and a doll, between the animate and inanimate—in other words, between life and death? Will reason win out over randomness? Will God get the travellers to Sweden? What can we know for sure?

It seems barely surprising that these concerns should have been traced back to, or posthumously inserted into, the life of Descartes, who is often referred to as the father of modern philosophy. They are philosophical problems (philosophy, until the nineteenth century, included all branches of science: mechanics, astronomy, botany, chemistry, anatomy, and so on), but they were relevant to everyone. Descartes's contemporaries and, more particularly, his immediate successors were moving from an age inhabited by alchemists and charlatans to one in which science was to be made transparent and accessible to all. A story is told about a Dutch cobbler who was teaching himself mathematics and wanted to discuss Descartes's method with him. Twice he visited the philosopher, and twice he was turned away by servants, who looked at his scruffy clothes and assumed he was a beggar. He rejected their master's offer of money, insisting that he only wanted "to speak of philosophy." On the third visit, Descartes welcomed him amongst his friends, and the cobbler, according to one of Descartes's biographers, "became one of the foremost astronomers of this [the seventeenth] century." It was also said of Descartes that he entertained the sick with mathematics.

The shift from exclusive knowledge and dark quackery to universal enlightenment was, however, an uneasy one. There was an abundance, in the eighteenth century, of manuals destined to train "ordinary minds" in the ways of physics and other related subjects. They had titles like "Philosophical Amusements" and "Mathematical Recreations"; they were meant both for pleasure and education, or education as pleasure. But although the Enlightenment project was to remove the veil from what the charlatans had previously peddled, the contents of these manuals were still on occasion called magic—and the general public, one imagines, must have found it hard to distinguish between sorcery and science.

Descartes had laid the foundations for one of the central ideas of that period: the notion, taken up by anatomists and philosophers alike, that man is a machine, and can only be understood as such. You could say that androids were a crucial part of Descartes's thought—his *Treatise on Man*, which was published after his death, is founded on a comparison between a human being and a hypothetical "statue or machine," which operates like a clock or a hydraulic fountain. He had already put forward a "beast-machine" hypothesis, in which he argued that animals were machines, made up of mere matter, and that all of their faculties could be explained by mechanical means. The difference between beasts and men, he said, was that humans possessed a "rational soul," whereas animals were incapable of reasoned thought (the *cogito*, "I think therefore I am," sets out what separates us from matter). However, the idea that the soul was the source of human life was to become very contentious, and the atheist philosophers of the eighteenth century stretched Descartes's beast-machine premise to include human beings as well. It was even suggested that Descartes had meant to say this all along, but had been too

7

afraid: his hypothetical moving statue was not an analogy, a later thinker said, but plainly a description of ourselves. His masking rhetoric was just a clever "ruse," "to get the theologians to swallow a poison."

So the man most famous for the dictum "I think therefore I am" was as interested in the way bodies worked as he was in the function of the mind (whilst Descartes was conducting his own anatomical investigations, the local butcher would deliver animal corpses for him to dissect at home). Neither the idea that men are machines, nor, conversely, the machines that were constructed to look like men, can be properly understood without him.

Jaquet-Droz's writing automaton in Neuchâtel is known to have scrawled, on some occasions, the words "I think therefore I am." At other times, it has written a more ironic tribute: "I do not think . . . do I therefore not exist?" It's a perfect riddle, of the kind many automata conjure up. The writer, a mere machine, is able to declare that it cannot think. Clearly, however, it does exist: and if it is able to communicate the fact that it cannot think, is it possible that it can think after all? Might the machine be lying? What is the difference between the automaton that writes "I do not think" and a person who, having lost the power of speech, is obliged to write that sentiment or its opposite on paper?

In this context, what the fable about the ship finally represents is the throwing overboard of one of Descartes's great contributions to philosophy, anatomy, and mechanics. Science was cast out to sea.

Indeed, for the supporters of these ideas, there was much to fear. The power of the church was oppressive, and would remain so for some time. Descartes had originally written *The World*, of which the *Treatise on Man* is the second part, in the early 1630s, but he had abandoned it on hearing of the fate of Galileo, who had been put under house arrest by the Roman Inquisition after supporting the claim that the earth moved around the sun. What would have been Descartes's first book became his last. He was not an atheist, but some of his ideas were seen as such, and he understandably feared the fickle interpretations of the church. He wrote to a friend of Galileo's conviction,

> I was so surprised by this that I nearly decided to burn all my papers, or at least let no one see them. For I couldn't imagine that he—an Italian and, I believe, in favour with the Pope—could have been made a criminal, just because he tried, as he certainly did, to establish that the earth moves . . . I must admit that if this view is false, then so too are the entire foundations of my philosophy, for it can be demonstrated from them quite clearly. And it is such an integral part of my treatise that I couldn't remove it without making the whole work defective. But for all that, I wouldn't want to publish a discourse which had a single word that the Church disapproved of; so I prefer to suppress it rather than publish it in a mutilated form.

No matter how Descartes tried to appease the devout, however, the opposition between philosophy and religion was set. An eighteenth-century nobleman, speaking both of that philosopher's accessibility and the stubbornness of the monks, commented with chauvinistic wit that "fifteen years after the printing of Descartes' works, ladies reasoned much more sensibly in metaphysics than three-fourths of the nation's theologians."

Hence Descartes's careful insistence that the machine in his treatise is not a man, but only "a statue or machine . . . which God forms with the explicit intention of making it as much as possible like us." This machine is composed of a body and a soul: in the *Treatise on Man* he describes the body without the soul, and intended to describe the soul separately; but since this latter part of the treatise has been lost, what we are left with is a mechanical interpretation of everything in us except reason. And—though this conclusion may not have been intended—reason seems barely necessary, since not only do our lungs work like bellows and our blood flow as in a hydraulic system, but our memory, dreams, sleep, passions, hunger, pain, dizziness, and sneezes can all be accounted for mechanically. The treatise is a philosophical proposition stated in the language of medicine, an anatomical map of our insides, a description of the functions of human nature as if they were the various, linked junctures of a pinball machine. Descartes writes in conclusion: "I desire . . . that you should consider that these functions follow in this machine simply from the disposition of the organs as wholly naturally as the movements in a clock or other automaton follow from the disposition of its counterweights and wheels."

Mechanistic philosophy found a number of supporters, but the most radical and most openly atheistic upholder of the man-machine thesis was an eighteenth-century physician named Julien Offroy de La Mettrie. Curiously, La Mettrie had intended, early on in his life, to enter the church. He studied philosophy and natural science at a distinguished school (also attended by the future editor of the *Encyclopédie*, Denis Diderot), which, during the time he was there, had begun to teach the works of Descartes, until then banned from most curricula.

A family friend advised him to go into medicine, and persuaded his father to accept this more lucrative alternative to theology. La Mettrie went to Holland to study under the physician Herman Boerhaave, who laid a great deal of emphasis on clinical instruction and deduction from practical experiment. Boerhaave aimed to interpret medicine according to the laws of mechanics: when it came to understanding the function of a particular organ, he wrote, it was the mechanicians whose "oracles should be consulted." Boerhaave was to have a lasting influence on La Mettrie, who later translated many of his teacher's works into French.

La Mettrie went on to set up a local practice in Brittany; he wrote medical treatises on vertigo and venereal disease, and was employed as a doctor to the French national guards. After writing a controversial, mechanistic treatise entitled *The Natural History of the Soul*, he lost his job, and all copies of the book were condemned to be burned by the public hangman. From then on, it seems, La Mettrie built up quite a collection of enemies: he had alienated the theologians, then he satirized other doctors; he was ostentatiously hedonistic, and wrote books on laughter and sexual pleasure, all of which behaviour caused further offence.

He was forced to flee to Holland to publish his next book, even though it was published anonymously. *L'Homme machine* (literally, "The Man Machine," but translated as *Man a Machine*), La Mettrie's most famous work, appeared in Leyden in 1747, and the publisher was immediately forced by the church to deliver up all copies for burning. As soon as the author's identity was suspected, La Mettrie had to escape once again, this time to Prussia, where he was welcomed and supported by Frederick the Great. Frederick, who wrote a eulogy to him after his death, made La Mettrie his personal physician, appointed him to the Royal Academy of Sciences, and was thought to have shown him a degree of favouritism that made him the envy of others in the King's circle.

In all this time La Mettrie's decadence showed no signs of letting up; his excessive gaiety, he once implied, masked a certain amount of sorrow. One night in 1751, after consuming a large amount of pheasant and truffle pâté at the home of an ambassador he had recently cured, La Mettrie fell ill, and died some days later. His sudden demise—as if from hedonism or gluttony—was met with triumph by his pious opponents, who thought it a fitting punishment for his life. So the man who was one of the key figures in the Enlightenment became, for more than a century, a joke, known only for the manner of his death.

L'Homme machine, whose influence on the materialist *philosophes* has since been given its due weight, took certain crucial steps beyond the position espoused by Descartes. La Mettrie attempted to show how like beasts men are, comparing humans with apes, and even suggesting that apes might be taught to speak. According to that principle, the idea that beasts are machines ought also to apply to men.

Other parts of his argument were the result, as Descartes's

had been, of important advances in contemporary science. William Harvey had already proved in the 1620s that blood circulated in the body; Descartes was much indebted to this discovery. By the time La Mettrie came to write *L'Homme machine,* a scientist named Abraham Trembley had recently found that the freshwater polyp, long classed as a plant rather than an animal, had the ability to regenerate itself when divided: it would, without intercourse, turn into as many polyps as there were parts. La Mettrie relied on this information to show that life is a property of matter, not dependent on a separate entity called a soul.

Another discovery that became central to La Mettrie's thesis was what is known as the principle of irritability. Albrecht Haller had been able to show that muscles move of their own accord—they respond individually if directly stimulated, rather than being reliant on "animal spirits," which Descartes believed in, and which were thought to cause the entire muscular system to move. Though Descartes had understood the body as a mechanically moving machine, he did not see it as a self-moving machine; that is, he did not think, as La Mettrie did, that the human body contained within it the principle of its own life. Through Haller's discovery, La Mettrie was able to argue that man was an automaton, or, as he put it, a "self-winding machine, a living representation of perpetual motion."

The soul, he wrote, was nothing but "an empty word to which no idea corresponds"; it should never be used to mean a source of life, but only to mean the mind, "to name the part in us that thinks . . . Where is the seat of this inborn force in our bodies?" he went on.

Clearly it resides in ... the very substance of the parts, excluding the veins, arteries, and nerves, in short, the organization of the entire body, and ... consequently, each part contains in itself springs whose forces are proportioned to its needs. Let us consider the details of these springs of the human machine. Their actions cause all natural, automatic, vital, and animal movements. Does the body not leap back mechanically in terror when one comes upon an unexpected precipice? And do the eyelids not close automatically at the threat of a blow? ... Does the stomach not heave automatically when irritated by poison, a dose of opium and all emetics? ... Do the lungs not automatically work continually like bellows?

The words "automatic" and "mechanical" are fundamental here, loaded terms that carried within them an inflammatory philosophical debate, pitting materialists against theologians. Although La Mettrie was a physician, and aimed to use medical experience in order to prove his arguments, he was also greatly influenced by progress in other branches of science, and mechanics in particular. Clocks, telescopes, microscopes, and various measuring devices were becoming ever more sophisticated. Descartes had already referred in his treatise to the mechanical fountain Tomaso Francini had built at the Royal Gardens in St. Germain en Laye in about 1600, as he suggested how the body was infused with animal spirits. In *L'Homme machine* La Mettrie, doing away with the spirits, mentions Huyghens's planetarium, a moving model of the solar system, and the automata of Jacques de Vaucanson.

The Blood of an Android

On the subject of these creations, La Mettrie points out: "If more instruments, wheelwork and springs are required to show the movements of the planets than to mark and repeat the hours, if Vaucanson needed more art to make his *flute player* than his *duck*, he would need even more to make a *talker*, which can no longer be regarded as impossible, particularly in the hands of a new Prometheus."

What La Mettrie is proposing is that the more human the machine, the more complicated its mechanism. Humans may contain more springs and wheels than animals, say, but they do not contain anything other than springs and wheels; and the closer eminent mechanicians like Huyghens and Vaucanson came to reproducing the world and the body in clockwork, the more they seemed to prove that man and the world were little else. "However greatly these proud and vain beings desire to exalt themselves," La Mettrie wrote disdainfully of his fellow men, "they are at bottom only animals, perpendicularly crawling machines."

The mechanical, or artificially animated, human became a common trope amongst La Mettrie's compatriots; and although La Mettrie set it down to most dramatic effect, the idea was already in the air. Aram Vartanian, who wrote an introduction to *L'Homme machine*, quotes an anonymous article written in a widely circulated journal, published three years before La Mettrie's book:

> If you had never seen anything but mounds of lead, pieces of marble, stones and pebbles, and you were presented with a beautiful wind-up watch and little automata that spoke, sang, played the flute, ate and

drank, such as those which dextrous artists now know how to make, what would you think of them, how would you judge them, before you examined the springs that made them move? Would you not be led to believe that they had a soul like your own, or at least like that of an animal, and would you not be led to wager that this very soul was the cause of their *activity*? Nevertheless, there is nothing in these automata but matter.

Diderot took up these thoughts in an unfinished treatise: "What difference is there between a sensitive and living watch, and a watch made of gold, iron, silver or copper?" The philosopher Etienne Bonnot de Condillac wrote a book arguing that the senses were a form of knowledge; his central image is a marble statue coming to life, sense by sense, from smell to touch. And the Greek myth of Pygmalion and Galatea underwent an extraordinary renaissance in the eighteenth century: Diderot, Rousseau, Voltaire, Rameau, and Deslandes were amongst the many writers who produced modern versions of the story, in which a sculptor falls in love with his statue and is rewarded when Venus brings Galatea to life.

Androids, or automata in human form, brought the figures of the clock and the statue together. They influenced these thinkers, as is clear from La Mettrie's mention of them, and they also seemed to be concrete proof of what the philosophers proposed. Men understood as machines and machines built to resemble men went hand in hand—it hardly mattered which had come first. Androids were more than mere curiosities: they were the embodiment of a daring idea about the self. Or, as Simon Schaffer has memorably put it, they were "both arguments and entertainments." It was the golden age of the philosophical toy.

The reigning genius of this mechanical world was Jacques de Vaucanson. Though Vaucanson achieved most notoriety as the producer of a high-society spectacle, in which an android played the flute and a mechanical duck was seen to digest its food, there seems little doubt that he was keen to align himself with philosophers and anatomists, and sought to contribute to the debates of his time. His magnificent creations were admired by audiences all over Europe; they were praised by kings and applauded by scientists. During the hundred years they were in circulation, they were exhibited in fashionable showrooms, at carnivalesque fairgrounds, and in private cabinets of curiosities: Vaucanson's automata crossed many boundaries, and might be seen as emblems of the Enlightenment, full of instructive clockwork and covered in gold, the legacy of mathematicians and alchemists. Both Voltaire and La Mettrie labelled him a "new Prometheus." Like the Greek Titan, he had the power, it seemed, to create life, to fashion men out of new materials. He was, as his biographers André Doyon and Lucien Liaigre point out, an early cybernetician, and though his career would later stray from this Promethean path, his wildest and most secret ambitions were to remain in the realm of artificial life.

Jacques de Vaucanson's earliest mechanical influences came from the church. He was the youngest of ten children (born in Grenoble in 1709), and his Catholic mother would take him with her every time she went to confession. While his mother was with the priest, Jacques stared at the clock in the adjoining room. Soon he had carefully calculated and memorized its mechanism, and was able to build a perfect copy of it at home. His father, a master glovemaker, died when Jacques was seven, and the boy was sent away to be schooled at a monastery, where he arrived clutching a metal box. He didn't get on with the other boys, and couldn't concentrate on his lessons, which he spent drawing strange lines on pieces of paper. Eventually, the Father Superior was forced to open the box. He found wheels and cogs and tools, next to the unfinished hull of a model boat. When confronted, Jacques refused to do any studying until he could make his boat cross the school pond. He was locked in a room for two days as punishment, but he spent the time making drawings so exceptional that the mathematics teacher, who was later to be lauded by the Royal Academy of Sciences, decided to help him.

Of course, a story exists about the youthful genius of all famous men. What is curious here is that all of Vaucanson's early efforts as a mechanician were connected in some way to religion. The clock was seen at confession; the maths teacher was a monk. He went on to be taught by Jesuits, and, on leaving school, became a novice in the religious order of the Minimes in Lyon. This was the only way, he thought, that he would be able to pursue his scientific study, given the limited finances of his widowed mother. At the time, there was no reason to suspect anything to the contrary; many members of the Royal Academy of Sciences carried the word "Abbot" before their name, and some of the cleverest people in the country were to be found in religious institutions. Indeed, Vaucanson was given his own workshop in Lyon, and a grant from a nobleman to construct a set of machines; but his talents were only encouraged up to a certain point. In 1727, to celebrate the visit of one of the heads of the Minimes, he decided to make some androids, which would serve dinner and clear the tables. The visitor appeared to be pleased with the automata, but declared afterwards that he thought Vaucanson's tendencies "profane," and ordered that his workshop be destroyed.

Why was some clockwork tolerated when this was not? The so-called profanity was connected, surely, to the shape, or the function, of the automatic waiters. What was dangerous, then, was not the element of mechanism, but the element of man in these constructions. To liken man to a machine was unacceptable. Expecting the support of his fellow monks, Vaucanson had unwittingly stepped into the very ground that La Mettrie was to know, some years later, would make him a wanted man. From this point on, Vaucanson realized he was involved in a risky business. He went home to Grenoble, and, offering the excuse of an "unmentionable illness," pleaded with the Bishop to be withdrawn from the order. As soon as he was free, he ran away to Paris.

However elegant or genteel his inventions might seem to us now, it should not be forgotten that Vaucanson, along with many contemporary philosophers and surgeons, was treading a fine and dangerous line. As his automaton-making career progressed, he became as distanced from his former calling as it was possible to be. Many years later, Vaucanson found himself in Lyon once again; this time, however, he was there not as part of the monastery, but in order to introduce new technology that would transform the manufacture of silk. The silk workers rioted, and, under threat of terrible violence, Vaucanson was forced to flee the city. In his search for a disguise, he lighted on an outfit he had once worn in good faith, but which had now become the emblem of his philosophical opposite. Vaucanson escaped by night, undetected: no one could have guessed that an atheist mechanician was hiding beneath the habit of a Minime monk.

Little is known about Vaucanson's activities around the time he left for Paris. It is thought that he attended classes in anatomy and medicine at the Jardins du Roi (the Royal Gardens), subjects that he continued to study on a visit to Rouen. In Rouen he probably met Claude-Nicolas Le Cat, who had just been made head surgeon at the hospital there. Le Cat, as we shall see, was involved in the construction of an artificial man, and it's possible that Vaucanson made an automaton under his tutelage. One of Vaucanson's early machines, which represented different animals in motion, was powered by fire and water, and he had soon produced enough work to go on an exhibition tour of Brittany. In Tours he met one of his main financial backers, and returned to Paris with enough money to dress in floral garments and carry a sword—in short, to gain a gentlemanly entry into high society. He met Voltaire and other philosophers; he was introduced to the future finance minister, Bertin; he came into contact with the best musicians in France. All of these people would support him in one way or another in the years to come.

Just as he was preparing to construct the automaton he had been sponsored to make, however, Vaucanson fell seriously ill. He was bedridden for four months, and could not eat for half of that time. Surgery sank him deeper into debt. In his delirium, he dreamed up an android that could play the flute, an android in the shape of a famous marble statue by the royal sculptor Antoine Coysevox, then on display in the Tuileries Gardens. He rose from his bed and drew up designs for every part, handing them out as he went along to various craftsmen and clockmakers. As soon as the pieces were joined together, the automaton could be heard to play the flute, as perfectly as any human being. It was as if the marble statue had come to life.

Vaucanson rented an elegant showroom in a grand mansion in the centre of Paris, the Hôtel de Longueville. The walls were covered in mirrored panels and gilded ornaments, with frescoes over the doorways; two six-panelled windows gave out on to a central courtyard. The room was called the Salon des Quatre Saisons, and Vaucanson, taking on airs which would guarantee him a fashionable audience, began to have his post sent there.

The Automaton Flute Player was first exhibited in that room on 11 February 1738. The price of entry was three livres, a week's wages for a manual labourer. Vaucanson demonstrated the object himself, to groups of ten to fifteen people at a time. The show was a huge success: by April that year, there were seventy-five visitors a day, and the Academy of Sciences, who had initially responded half-heartedly to what they assumed was yet another piece of trickery, had given their full approval. As if to confirm its lasting success, a lengthy description of the Flute Player's mechanism appeared over several pages in the first volume of Diderot and d'Alembert's *Encyclopédie,* under the entry *"androïde,"* defined as: "an automaton in human form, which, by means of certain well-positioned springs, etc. performs certain functions which externally resemble those of man."

The figure was made of wood, and painted white to look like Coysevox's marble. It was life-size—five and a half feet tall—and was supported by a large pedestal measuring four and a half feet in height and three and a half in width. The flute, as Vaucanson had learned from his musical acquaintances, was considered one of the hardest instruments to play in tune—notes are produced not just by fingers and breath but by varying amounts of air blown into the flute, and different shapings of the lips. He had set himself an apparently impossible task, and emerged with a machine that could play twelve different melodies. At first, spectators believed the sounds must be coming ready formed from inside the figure—they had seen musical toys before, but never had an automaton actually played an instrument as a human being would. The virtue of this Flute Player, and the reason it seemed an ideal Enlightenment device, was that Vaucanson had arrived at those sounds by mimicking the very means by which a man would make them. There was a mechanism to correspond to every muscle.

The account of it he gave to the members of the Academy of Sciences, and which he subsequently published, illustrated with an engraving by the well-known book-illustrator Gravelot, was broken down into two sections. First, he spoke about his investigations of the German flute itself, and exactly which muscles were required to play it, offering an anatomico-musical analysis of what it took to produce different notes, to jump an octave, to pause, to alter the pace, to play loud or soft. Then he explained, in almost dizzying detail, the mechanism

he had created for his artificial man. The figure was constructed so that it could be opened to show its secret strings as
Vaucanson spoke. "The entire mechanism," he wrote, "is to be
seen uncovered, since my intention is to demonstrate, rather
than simply to show a machine."

The mechanism, needless to say, was extremely complicated—full of screws and pivots, barrels and bars—and I shall
explain it only briefly here. The heaviest elements were to be
found in the front and back parts of the pedestal. In the front,
several wheels were set in motion by a weight, which in
turn carried around a steel axle attached to cranks, which
were attached to six bellows. In the back, there was a series of
different-sized pulleys, which were connected to three more
bellows. The strings on some of the pulleys led to levers and
valves, which eliminated noise or excessive movement when
air was pumped through. The nine bellows were attached to
three separate pipes that led into the chest of the figure. Each
set of three bellows was attached to a different weight to give
out varying degrees of air, and then all pipes joined into a single one that was equivalent to a trachea, continuing up through
the throat, and widening to form the cavity of the mouth. The
mouth, though the smallest part, contained the most intricate
apparatus. The lips, which bore upon the hole of the flute,
could open more or less depending on the amount of air that
was to be passed into the instrument, and they could move
backwards or forwards. For each of these four movements
there was a separate mechanism. Inside the mouth was a moveable metal tongue, which governed the air let through and created pauses. There were four levers to operate the tongue and
to modify the wind. Other devices ruled the player's fingers;

all of these motive forces were hidden in the pedestal, and were made up of a cylinder, a key-frame, fifteen levers, and numerous steel wires and chains.

Vaucanson's intention, he said, was to compare these motions and effects with "those of a living person," and it is striking that his description bears a strong resemblance to the point-by-point map of the human body offered by Descartes a century earlier. The bellows, the clockwork, the pipes are there in both. Vaucanson's levers and valves are not so far from Descartes's tubes and membranes, and the latter likened the flow of animal spirits to that of a wind passing through the body—as it does through the Flute Player. Descartes's description is full of analogies with mechanical or musical contraptions (a hydraulic fountain, a church organ), just as Vaucanson's is a comparison with the workings of the human body. In other words, reading the anatomical description of the "human machine" and reading the mechanical explanation of an artificial man are very similar experiences.

Vaucanson's android inevitably raised questions about what it meant to be human. It seems that its primary uncanny effect stemmed from the fact that it operated by breathing. Clearly, almost any other instrument, requiring only physical pressure in order to produce a sound, could be played simply by clockwork (there had been automated bell-ringers and lute-players, and later that century Henri-Louis Jaquet-Droz was to construct his famous Musical Lady, which played the key-board). This automaton *breathed*. Even though the art of mechanics was sophisticated enough by then to make a machine perform many other movements, and even though Vaucanson unveiled the fact that this breath was created by bellows, the very act of breathing, seen in an inanimate figure, continued to cause a stir well into the following century. The first mechanized waxwork in Madame Tussaud's was Sleeping Beauty, who was said to have been modelled on Louis XV's mistress Mme. du Barry, and whose sole mechanical feature was a heaving chest. In 1833 a breathing model of the late Napoleon did the rounds of London showrooms and was said to be "the astonishment of the medical world."

The Flute Player, however, also raised questions about perfection, and what was understood by it. A year later Vaucanson exhibited a figure that played a pipe and drum. It received little attention, partly because it was accompanied by a much more memorable automaton, and partly because it was so similar to the Flute Player people had already seen. But there was a crucial difference between the flute- and pipe-players: the pipe was played so fast that it exceeded the speed any living person could achieve (only a mechanical tongue could move that quickly), and it was also the instrument known to be most exhausting to the human chest—the android, however, never got tired. So the Pipe Player embodied one idea of perfection, the idea that humans were messy, imperfect, fallible, and that a perfect machine would correct those flaws, improve on humanity. The Flute Player, on the other hand, was meant to approach humanity as much as possible—perfection, in this case, being as close a replica of human imperfection as there could be. The evidence for this lies in a small problem Vaucanson encountered as he was making the android.

He had designed the seven levers corresponding to the fingers, and he had ensured that there was the appropriate range of angles for the mouth; but although the actions were all correct, and the passage of air was under control, the sound was not quite right. Vaucanson discovered that wooden fingers could not play a metal flute the way a man or woman could: the difficulty was that the machine was just not soft enough. Vaucanson looked around for a material that would accurately simulate the effect, and he found it. The glovemaker's son covered his android's fingers in skin.

It is possible that the material was leather, since the word *peau* in French makes no distinction between animal and human skin; but Vaucanson certainly never specified one or the other, and in any case, the point is the same: pure mechanics were not enough, and Vaucanson had to import organic matter into his dead creation. It's as if, in trying to come close to humanity, Vaucanson had crossed a line—a machine could not be truly like a man unless it borrowed from a man, unless it arrived at its mechanical concert dressed in skin. From this perspective, robbing graves to make a monster seems a small enough step, and the Flute Player leaves the theologians some room for argument: as a later commentator sarcastically put it, "What a shame the mechanician stopped so soon, when he could have gone ahead and given his machine a soul!"

In 1739, when attendance at the exhibition was flagging, Vaucanson added to the Flute Player two other machines, which he had, in all likelihood, constructed earlier and kept in reserve. One was the pipe-and-drum figure, and the other was a mechanical duck. If the Flute Player imitated life with human flaws and particles, then the duck went even further towards reproducing those parts in us for which a machine would have no need—because what was remarkable about this duck was that it ate food out of the exhibitor's hand, swallowed it, digested it, and excreted it, all before an audience. It became Vaucanson's most famous creation; without the shitting duck, Voltaire commented wryly, there would be nothing to remind us of the glory of France.

It was made of gold-plated copper, but it was the same size as a living duck, and moved just like one. Aside from its main digesting function, it could drink, muddle the water with its beak, quack, rise, and settle back on its legs, and, spectators were amazed to see, it swallowed food with a quick, realistic gulping action in its flexible neck. In a single wing alone, it was later revealed, there were more than 400 articulated parts. Vaucanson added an appendix about the duck to his "Memoir" of the Flute Player. It took the form of a letter to his friend, the influential journalist Abbé Desfontaines and, once again, was very precisely anatomical. One part of it described the duck's wings at great length, in order to give an example of the careful study involved in the construction of the machine:

I do not believe the Anatomists can find anything wanting in the construction of its wings. Not only has every bone been imitated, but also all the Apophyses or Eminences of each bone. They are regularly observed as well as the different joints: the bending, the cavities, and the three bones of the wing are very distinct. The first, which is the *Humerus*, has its motion of rotation every way with the bone that performs the office of *Omoplat, Scapula* or *Shoulder-blade;* the second bone, which is the *Cubitus* of the

wing, has its motion with the *Humerus* by a joint
which the Anatomists call *Ginglymus;* the third, which
is the *Radius,* turns in a cavity of the *Humerus,* and is
fastened by its other ends to the little end of the wing,
just as in the animal. The inspection of the machine
will better show that Nature has been justly imitated
than a longer detail, which would only be an anatomical description of a wing.

The description was so elaborate, and the final ironic throwaway so effortless, that two people who handled the duck later
in its life were led to believe, independently of one another,
that Vaucanson had perfected the rhetoric of a magician. The
duck did not function as he said it did, they thought, and his
casual talk had served to obscure this fact.

The rest of Vaucanson's letter read like a brief treatise on
the nature of digestion. There was some debate at the time
over whether the process of digesting food involved "trituration" (grinding) or dissolution, with gastric juices, or a combination of the two. By presenting his duck to the French public
Vaucanson, though not a doctor or a philosopher, was actively
contributing to that debate. "Food is digested in its stomach,"
he wrote, "as it is in real animals, by dissolution, and not by
trituration, as many natural philosophers contend." He was
clear, however, about what his artificial duck could and
couldn't do: "I do not pretend to offer this digestion as a perfect digestion, capable of producing blood and nutritional elements for the animal's continuing health; I believe it would be
churlish to reproach me for that." By being so specific, Vaucanson was guaranteeing that his audiences would believe that

the duck was in fact performing these allegedly limited, though clearly extraordinary, feats.

He went on to give details of the duck's insides: not only was the grain, once swallowed, conducted via tubes to the animal's stomach, but Vaucanson had also had to install a "chemical laboratory" to decompose it. It passed from there into the "bowels, then to the anus, where there is a sphincter which permits it to emerge." Elsewhere in the tract he described how the digested matter was "driven away at Pleasure through circumvolutions of pipes," leading an American observer to make a speculative drawing of the duck's insides, often supposed to have been drawn by Vaucanson himself. The drawing reminds us how strange it was for Vaucanson to create these "circumvolutions," which look very much like intestines, when he could just as easily have laid down a straight pipe to carry the food to the other end of the duck. The fact that he chose to emulate the form when he need only have copied the function implies that there may have been some sort of projection, or identification, involved.

Of course, it was a peculiar project altogether. As Voltaire must have meant, France now had as its glorious mascot a golden creature that was famous for its excrement—ingenious, yes, and full of mechanical marvel, but also a reminder that civilization could not be separated from its waste. There was mess involved in even the least fleshly pursuits. Indeed, the most immediate wonder of Vaucanson's duck was that it performed functions that would never be expected of a machine; it was beyond a machine, it was a highly skilled joke. Had the duck been an artificial defecating man, there would no doubt have been a more complicated, less rapturous response; but, aside from his contribution to fashionable debates, the question of why Vaucanson should have wanted to manufacture such an artefact remains.

Vaucanson, it must be said, was a man much preoccupied by the state of his body. We know that he designed an automaton whilst plagued by an illness that had prevented him from eating. Though details of his condition would otherwise be superfluous, it is worth noting, given Vaucanson's language in describing his duck, that he was suffering from a fistula of the anus. It was a painful affair, which the medical profession took very seriously—Louis XIV had been the victim of one as he was dying, and the *Encyclopédie* contains, in its section on surgery, diagrams for operating on it. The mechanician's particular mention of the bowels, anus, and sphincter of the duck—parts audiences may have preferred to imagine for themselves—might be seen as a reflection of his own personal preoccupations.

This notion is purely speculative, but it is supported by two other occasions when Vaucanson used his apparently weak physical health as an excuse. The first was when he wanted to leave the Minimes: he told the Bishop of Grenoble that the vegetarian diet the order was committed to was not good for his delicate digestion (presumably, his "unmentionable illness," if real, was related to his digestive system). The second was after the duck had been exhibited. In 1740, Frederick the Great, recently having acceded to the Prussian throne, and following the suggestion of Voltaire, invited Vaucanson to join his Academy of Sciences, as he later did La Mettrie. Vaucanson wanted to stay in France, but he would not let patriotism be his only answer. He had to stay, he told Frederick, because of his fragile health—he did not go into detail.

Whether or not Vaucanson's health really was that feeble is uncertain. What is clear is that his immediate response, when looking for a way out, was to plead illness, to make reference to his malfunctioning body. His machines, on the other hand, were flawless, superhuman things. They were, perhaps, imaginary prostheses: unlike its creator, one of them had a digestive system that ran like clockwork.

By 1741 Vaucanson had had enough of his automata. He wanted them to be shown in England, where there was a substantial audience for mechanical exhibitions, but he did not want to take them there himself. He had never seen himself as a mere entertainer, and in any case by then he had been given another, rather grand job. So he packed off his three machines with three Lyonnais businessmen, who paid over the odds for the privilege, and who later bought the automata outright. They disappeared from their inventor's view, and embarked on a new stage of their mechanical lives.

Since the act of telling time seemed to emulate a certain kind of intelligence, the clock became the ideal machine for the Age of Reason. But it also contained an ironic catch. As Aram Vartanian observes in his book on La Mettrie, "If on the one hand the clock is able to tell time, it remains on the other totally oblivious to time; and this very resistance to an inner shaping by virtue of duration is, in fact, what permits it to perform its task so admirably." Men are mortal, clocks are not. Vaucanson aged even as he constructed his ageless creations. They passed from one owner to another, and survived the death of those who possessed them; and so the history of Vaucanson's automata after their sale is one of disappearances and detective work, of improbable travels and sudden resurrections. The Flute Player left few traces, but the duck appears to have risen now and then, like a clockwork phoenix.

The ringleader of the Lyonnais trio was a man named Dumoulin, a perfumier and glovemaker like Vaucanson's father. He travelled with the machines throughout England and Holland, then back through France and Germany. He tried to sell them in 1754 to a relative of Frederick the Great, but the buyer couldn't find the cash. So Dumoulin pawned them in Nuremberg, and left for Russia, where he exhibited other curiosities. There are traces of his shows in St. Petersburg and Moscow, and though he was employed as a mechanic by the University of Moscow, he was sacked after six years, for incompetence. He died in St. Petersburg in 1781, leaving Vaucanson's automata still in the hands of the pawnbrokers.

At this point, they were seen, packed up in boxes and kept in an attic, by the German writer Christian Friedrich Nicolai, who published an account of his European travels in 1783. He reported that they had apparently been well preserved, though after "28 years of captivity" the cost of repairing them would be impossible to calculate. Before he left, Dumoulin had stored them in such a way as to make it impossible for anyone to exhibit them with any success. He had taken the two androids to pieces, and mixed up parts of the Flute Player with parts of the Pipe Player. The duck was intact, but Dumoulin had carefully positioned its internal chains in reverse, so that they would break if the duck was set in motion. Nevertheless, the mechanism of the duck was the easiest to see and, on closer inspection, Nicolai found that it did not digest its food at all. There was no "chemical laboratory," he revealed—the food was simply aspirated into the neck with the aid of bellows and tubes, and a separate substance made to look like the digested version was held at the ready in another compartment near the bird's rear end. This was "expelled at the desired moment by a piece of mechanism."

The result of Nicolai's report was that the machines were rescued by an extraordinary man whose father had told him about them when he was a boy. He had always wanted to own automata like them, he said, and when he was finally in possession of the very same objects he declared them to be "the greatest masterpieces of mechanics that humankind has ever created."

The man was Gottfried Christoph Beireis, doctor to the Duke of Brunswick, chair of Medicine at the University of Helmstadt, collector of curiosities and reputed master of alchemy. As a young medical student, Beireis had developed a taste for alchemical experiments and tricks of natural magic. He was an avid reader of a twenty-volume, late-eighteenth-century German work by Johann Christian Wiegleb, entitled *Instruction in Natural Magic, or, All Kinds of Amusing Tricks,* which mentioned Vaucanson's automata in its opening volume. E. T. A. Hoffmann also possessed a copy, and it was this book that contained the automaton-making instructions Hoffmann attempted to follow before he set out to write his stories "The Automata" and "The Sandman."

After finishing his studies, Beireis went on a long journey and returned extremely wealthy, leading his acquaintances to suppose that he had discovered how to turn base metal into gold. He did nothing to dispel these rumours, and developed a signature line in social magic tricks. On one occasion he was invited to dinner at the home of the Duke of Brunswick, where several dignitaries were in attendance. Beireis arrived wearing a bright-red coat, which he refused to remove. During the course of the meal, the coat turned completely black and fell to pieces on the floor. At that same moment, the Bishop of Hildesheim, who was sitting across the table, found that his wine had turned to vinegar.

Beireis was made Professor of Philosophy and Medicine at Helmstadt when he was still very young. He bought a large house with a garden, and filled it with musical instruments, rarities of natural history, jewels, paintings, anatomical exhibits, and mechanical objects. He was a gifted fencer, and quite a dandy; he wore early-eighteenth-century clothes, with diamond-encrusted buttons and buckles for special occasions, until his death in 1809. When he bought Vaucanson's automata, he wasted no time in dressing the two androids in gold and silver outfits.

In 1805, Goethe went to visit Beireis, lured by the legend of "the old wonder-worker," as he called him. His description is well worth quoting at length:

Hofrath Beireis, an eccentric, problematic man, already for many years notorious in so many respects, had been so often named to me; his neighbourhood, remarkable possessions, strange behaviour, and the secret brooding over all, so often described to me, that I could not but reproach myself with the fact that I had not seen with my own eyes, and in personal inter-course endeavoured to fathom, in a certain measure at least, this most singular personality, which seemed to point to an earlier, transitory epoch . . . Professor Wolf being in the same predicament in this respect with myself, we determined, knowing the man was at home, on undertaking a journey to the mysterious griffin who presided over extraordinary and scarcely conceivable treasures.

Goethe and his companions were greeted by a man who was "not tall, well-built and agile, the legends of his fencing skill may well be believed; an incredibly high and vaulted fore-head, quite out of proportion with the lower and finely con-tracted parts, indicated a man of singular spiritual force." They were shown his art collection—paintings by Titian, Raphael, Correggio, Rubens, Dürer—and saw, in the middle of a large hall devoted to natural history, a series of stuffed birds, "all eaten to pieces by moths, feathers and vermin lying heaped up on the stands." They found our artificial bird there too, in no better condition: "A great deal of his former posses-sions," Goethe reported,

the name and reputation of which were still fresh, we found in the most lamentable state. The Vaucansonian automatons were utterly paralysed. In an old garden-house sat the Flute Player in very unimposing clothes, but his playing days were past . . . A duck without feathers stood like a skeleton, still devoured the oats briskly enough, but had lost its powers of digestion.

With all this, however, Beireis was by no means put out, but spoke of these obsolete, half-wasted things with much complacency, with an air of much conse-quence, as if he thought that mechanism had since produced nothing new of greater importance.

On Beireis's death it became apparent that Napoleon had offered him money for the Vaucanson machines, but that Beireis would not let go of them. In 1810, his heirs offered to sell them back to France; in February that year there was some correspondence between various noblemen about this, inquir-ing whether Napoleon's interest still held. "Monsieur le Duc," Count Beugnot wrote to the Duke of Berg,

> The cabinet of Mr. Beireis contains amongst other curiosities the three automata of Jacques Vaucanson. His Majesty the Emperor and King wished to bring back to France these surprising productions which had left the country, and offered to buy them from professor Beireis six years ago.
>
> But this *savant*, renowned as much for the strangeness of his character as for the extent of his knowledge, turned down a proposal whose intention and dignity he was unable to appreciate.
>
> Count Brabeck . . . informs me that professor Beireis has recently died and that his heirs are willing to cede to the desire which his Majesty deigned to express, and to give him back the Vaucanson pieces. He adds that the professor has also left a collection of gold medals and some anatomical preparations.

I thought it my duty to report to your Excellency Count Brabeck's message. He is not as yet able to indicate what price professor Beireis's heirs have put on these items of his collection: but if his Imperial Majesty still desired to enrich France with their presence, his minister at the Cassel court could undertake the negotiations with the help of Monsieur de Brabeck. Of the three items proposed, the first . . . at least belongs to France. I believe I remember reading the description of it and its praise in the proceedings of the Academy of Sciences, and his Majesty will surely want to reconquer these monuments of French genius which have gone astray.

This time, however, it was Napoleon's turn to reject the offer, and the automata disappeared once again; only the duck has been heard of since. It was found twenty years later in the attic of another pawnbroker, by Georges Dietz, a theatrical impresario and exhibitor of automata. Dietz passed it on for repair to a famous Swiss clockmaker, Johann-Bartholomé Reichsteiner, who spent three and a half years working on the duck, which he said contained several thousand pieces. Reichsteiner concentrated so hard on repairing the automaton that he became quite ill, but eventually Dietz was able to exhibit it once more, at La Scala theatre in Milan in 1843. In the meantime, Reichsteiner, who had whilst studying Vaucanson's creation thought of a number of possible improvements, had built an alternative model. His version (and possibly Vaucanson's as well, after he had fixed it) was covered in real feathers over its gold and copper ones. And since Reichsteiner began to exhibit his own duck not long afterwards, there remains in the rest of this story some confusion about whether the duck in question was the original or a clever copy. At any rate, Dietz certainly took Vaucanson's duck to Paris in 1844 for the Exposition Universelle at the Palais Royal, where a wing fell out of order.

Also on show at the Exposition were the automata of a celebrated magician, Jean-Eugène Robert-Houdin (the conjuror from whom Houdini took his name). Robert-Houdin had made an automaton that could write and sketch, which was given a gold medal at the Exposition. Dietz asked him to repair the duck's broken wing, and the magician, delighted to have his hands on the famous creature, wrote about the occasion in his memoir. "To my great surprise," he reported gleefully of Vaucanson, "I found that the illustrious master had not been above resorting to a piece of artifice I would happily have incorporated in a conjuring trick." Robert-Houdin discovered what Nicolai had found sixty-three years earlier: the digestion had been faked, and the emitted substance was a premixed preparation of dyed green breadcrumbs, "pumped out and collected with great care on to a silver platter."

It should be borne in mind that many of the stories in Robert-Houdin's memoirs were rather far-fetched—as late as 1928, the authors of the most scholarly book on automata, Alfred Chapuis and Edouard Gélis, refused to believe Vaucanson could have cheated. Robert-Houdin's reaction, however, was to admire the late inventor even more: "Clearly, Vaucanson was not just my master in mechanics—I must also bow before his genius for trickery . . . Far from diminishing the high opinion I had of Vaucanson, this artifice, on the contrary, made me admire him doubly, for his knowledge and for his know-how."

Subsequent traces of the duck are scarce—a mechanical one was said to be in the possession of a certain Blaise Bontems, who manufactured singing birds, in 1863. Another rumour had it that the duck had met its end in Nijni Novgorod. In 1882 someone wrote a letter to a German newspaper claiming they had seen the duck in Krakow in the summer of 1879. He and his wife had been to see the cabinet of a Mr. Gassner, the letterwriter said. Gassner was exhibiting a collection of wax figures, antiques, and a blueprint of Thomas Edison's phonograph, which had been invented the year before, along with a duck he claimed had belonged to Vaucanson. They were delighted by the show; but a few days later, as they were sitting in a garden nearby, the man and his wife smelled burning. When they went to investigate, they found Gassner's museum razed to the ground—a small flame had made the waxworks catch fire, and the blaze had spread out of control. Amidst the ashes, he reported, they found a pair of misshapen metal wheels, "the pitiful remains of our glorious bird."

More recently, however, some mysterious photographs have come to light, and have been preserved in the archives of the Conservatoire National des Arts et Métiers in Paris (which was founded after the mechanician's death in order to house his remaining machines, and is situated on a street now called rue Vaucanson). They show a crude, featherless bird, made of spring-like windings of wire and perched on a huge wooden frame that contains a mechanism resembling a watermill. They are extraordinary views, reminiscent of the sorry skeleton

Goethe described, and give the impression that the figure is some sort of prehistoric find, better suited to a museum of natural history than to one of arts and crafts.

The provenance of the photos, however, and the true identity of what they depict, are still in question. They were found in a drawer by the museum's conservator in the 1930s, and they were marked "images of Vaucanson's duck, received from Dresden." There was no further explanation, though the archives possess a letter written by an Italian man to the museum's director in 1899, offering him not only the photos, but the machine as well (so if this were Vaucanson's duck, a different one must have been destroyed in Krakow). The director's reply was that the condition of the object was too poor, and the price proposed too high, for him to consider buying it. This is all the information we have. The present director of the museum does not believe the bird in the pictures is the original duck; Doyon and Liaigre, Vaucanson's biographers, believe it is. Either way, these photographs, the last fragments of possible evidence, tell their own story: Vaucanson's artificial beings broke free from their creator and developed an afterlife of their own; they were stripped back and rebuilt, seen through and newly admired—whether in truth or in legend, they continued to survive.

As soon as he had sold his automata, Vaucanson put all his energies into his new job. Louis XV had been a great admirer of the duck, and in 1741 he appointed Vaucanson Inspector of Silk Manufacture in the kingdom. Until then, there had been specific problems in the making of silk: although silkworms were bred in France, there was no effective system for collecting the cocoons or conserving them, and because the mills and wheels were mediocre, the weaving process made the resulting fabric faulty, so the French had resorted to importing the raw materials from Italy. Vaucanson made several lengthy trips to Lyon, the silk-making capital and home of his former monastery. By introducing new regulations and designing new looms (one of which became the basis for the better-known Jacquard loom), Vaucanson revolutionized the industrial process in France. He left the realm of fashionable curiosities, and concentrated on reforming the world of work. Although not strictly automata, these machines were in a sense prostheses—extensions of men—or substitutes for men. In the shift towards industrialization, labourers were valued less as machines were valued more, and Vaucanson played a significant role in the widespread replacement of men with their artificial counterparts.

In his funerary tribute to Vaucanson, the Enlightenment mathematician and philosopher Condorcet defined a mechanician as one who "sometimes applies a new motor to machines, and sometimes makes machines perform operations which were previously forced to be reliant on the intelligence of men; or he is one who knows how to obtain from machines the most perfect and abundant products." This, according to the silk workers of Lyon, was precisely Vaucanson's wrongdoing. They rebelled against his automatic loom by pelting him with stones in the street; they insisted that their skills were needed, that no machine could replace them. In retaliation, Vaucanson built a loom manned by a donkey, from which a baroque floral fabric was produced, in order to prove, as he said, that "a horse, an ox or an ass can make cloth more beautiful and much more perfect than the most able silk workers." This spiteful performance, surprising in the son of a craftsman, was the reverse of his golden duck: instead of producing excrement from a precious metal, he made luxurious silk emerge from the end of a live animal. The first was designed for man's entertainment; the second was meant to show man that he was dispensable.

The biographers Doyon and Liaigre blame the silk workers for stalling the march of progress, for France's Industrial Revolution lagging behind England's; and Condorcet comments melodramatically that "whoever wishes to bring new enlightenment to men must expect to be persecuted." The point of view of the workers seems to have been sidelined altogether in favour of a grand Enlightenment project. The *Encyclopédie* devoted sixteen pages (not including illustrations) to the making of silk and other stockings. "In what systems of metaphysics," it reads, "does one find more of intelligence, wisdom, consequence, than in machines for spinning gold or making stockings? . . . What demonstration of Mathematics is more complicated than the mechanism of certain clocks?" In the *Encyclopédie*'s illustrations, the men are secondary to the machinery. Vaucanson and his contemporaries contributed to a widespread sleight of hand: like wine into vinegar or base metal into gold, men were turned into machines. The new automata were not replicas, but real humans transformed. Throughout the next century, factory workers came to feel they had been reduced to the mechanical pieces they were in charge of producing, hour after hour and day after day.

Vaucanson's reforms were not limited to showy fits of pique. One of the first big organized strikes in French history erupted in response to new regulations he had put in place. In 1744, makers of silk fabric and stockings—labourers and journeymen, overseers of the manufacturing community, dye workers, carpenters, crochet workers, shopkeepers—and manufacturers of gold and silver cloth revolted. The uprising was so dramatic that the local government quickly took measures to appease the workers, but these were revoked by the King, who responded first by issuing fines, and then by prohibiting the workers, on pain of prison, from gathering in "cabarets, taverns, cafés and places of public games" in groups of more than four. The punishments became even more severe. A crochet worker by the name of Gaspard Jacquet was condemned to appear before the Palace and the Hôtel de Ville, holding a blazing torch, naked except for his shirt, with a sign around his neck that read "seditious crochet worker." After being fined, he was interrogated, on his knees, forced to reveal the names of his accomplices, and made to ask God for forgiveness. He was then hanged to death. Other strikers suffered the same fate; still more were imprisoned. Almost a year later, the King issued an amnesty; but the damage had been done. Vaucanson had tried to replace men with machines; men had died as a result, and he had been forced to escape violence under cover of night, disguised as a Minime monk.

Louis XV was twenty-nine when he saw Vaucanson's duck; he had been king since the age of five. Accounts of Louis's early life portray him as a lonely child—he had lost his great-grandfather, the Sun King, and his mother, father, and brother within the space of three years, and his rank separated him from most other children. Later in life he wrote in a letter of condolence to someone whose mother had died: "I have the misfortune never to have known what it is to lose a mother." Even when he had the entire palace to himself and ruled all of France, Louis and a few sons of noblemen constructed a kingdom in miniature: they played with his menagerie of toy animals, and, calling himself "Maréchal Duc Louis," he made dukes and marshals of all his friends, who became, collectively, the Order of the Pavilion, after the little outhouse where they sometimes played. (A single leather-covered toy elephant remains from this time, preserved in a German museum.) The fact that Louis had the stature of a child and the status of a king seems to have posed an interesting problem for artists of the period: in many paintings and engravings, in which he greets visitors and members of his court, he looks like a midget, or a dressed-up doll—a small thing of adult proportions, weighed down by gilded clothes and chains. In effect, he had no childhood, except by reducing his world to fit his size and, conversely, he was made into a symbolic, miniature object by others; it's not impossible to imagine that when Louis saw the duck, some part of a lost childhood had been found, in the form of a philosophical toy.

Louis was tutored by the Cardinal de Fleury, whom he later appointed as his regent, though he was old enough by then to rule the country on his own. Fleury taught him Latin, history, astronomy, geography; he took him to the Louvre, where the boy saw relief maps of his kingdom, to the Jardin des Plantes, where he met his great-grandfather's former doctor, and to the observatory, where he delighted in experiments involving magnets. Louis showed a particular interest in stories about voyages, and everything relating to them—globes, compasses, maps—and he developed a profound fascination for anatomy. Antoine Coysevox was one of his court sculptors, and Philidor, the renowned chess player, was retained as a musician. Louis was to become a great patron of astronomers and naturalists, and a supporter of numerous scientific expeditions. He loved, as is clear from his admiration for Vaucanson, all the mechanical inventions of the time. Although much has been written about the scientific progress made during the period of his reign, as Louis's most recent biographer, Michel Antoine, points out, the King's personal involvement and interest in it has been all but ignored. When Louis died, all sorts of mechanical apparatus were found in his cupboards and drawers: there were eight cases of mathematical instruments, four opera glasses, a portable barometer, a microscope, a telescope, a compass, and eleven watches.

Perhaps his most persistent interest, however, was the one on which his life depended. Louis was very close to his two chief surgeons, Mareschal and La Peyronie, the first of whom had attended his great-grandfather, and the second of whom was to remain with Louis for almost thirty years. Louis was nine when La Peyronie was appointed, and, from the time he was a boy, he would continually ask the surgeon questions about anatomy. Michel Antoine believes the King's concern derived from his parents' early death, and from his own persistently fragile health; clearly Louis's curiosity was precocious. Early on, La Peyronie instructed him in anatomy using artificial models of parts of the body, and dissected animals from the royal menagerie before him. Louis applauded these performances, and told everyone in his entourage about La Peyronie's cataract operations, which he had witnessed.

Acquaintances were sometimes shocked by the frequency with which Louis's conversation turned to matters of health, illness, and death. One reported an occasion when he had arrived at the palace, only for the King to announce straightaway that the foreign secretary had died. The foreign secretary had had, in his lifetime, a nervous tic that had made him grimace, and without pausing for his visitor to register the shock of the man's death, Louis said, "Do you know, he's been opened up, and it turns out he had a growth in his liver which made his tic more frequent as it increased in size?"

A famous surgeon later commented that Louis had wanted to know everything about "the very structure of the human machine," and that he could "hold a lengthy conversation with the most learned anatomist." He regularly attended the public demonstrations of apothecaries, was known to send delegations of surgeons to rural towns to stop the spread of epidemics, and he gave more noble titles to medical men during his reign than any other king had before him.

When Louis went to visit Vaucanson after seeing his artificial duck, he had a particular anatomical project in mind. Amazed by what the mechanician could do with the animal's insides, he asked Vaucanson if he could "execute the circulation of the blood in the same way." In other words, if he could make a man-machine.

It was several years before the Swiss physician Philippe Curtius began to model portraits in wax, and many more before he taught his protégée Marie Grosholtz, who would later become Madame Tussaud, but the practice of making anatomical models had been in place for some time. We know, for example, that Leonardo da Vinci (who once built an automaton in the shape of a lion) had succeeded in making casts of the brain by injecting its ventricles with wax, and Michelangelo later sculpted a number of muscle studies out of the same substance. Until the seventeenth century, these models had been the work of artists rather than anatomists, and they were used to teach art instead of medicine; but then the artists and the anatomists began to come together. The wax-modeller Gaetano Zumbo (whose work is still on display at the natural history museum of La Specola in Florence, along with the later, exceptional wax models of Felice Fontana) worked in collaboration with a surgeon named Guillaume Desnoues. When they went their separate ways, Desnoues set up a museum for the waxworks in Paris, which later moved to London, and on his death the models were sold at auction. At this point, the worlds of anatomy and spectacle collide. The models were kept in one place, in Covent Garden, on a sort of time-share basis: at certain times of day, they were used by surgeons to give anatomy lessons, and at others they were the subject of ghoulish tours for the curious. They went to Paris in 1729, and continued to be shown as a travelling exhibition until at least 1740.

In Vaucanson's time, an Italian couple called Anna and Giovanni Manzolini were making anatomical wax models with great success, and in France a woman named Marie Catherine Bihéron was widely admired among the *philosophes* and their correspondents. Mlle. Bihéron, whose *pièce de résistance* was the dissectible figure of a pregnant woman, was the first to have worked out a way in which her models could be taken apart, so that each section of the body, and how it connected to others, could be shown individually. Near the end of Mlle. Bihéron's life, Catherine the Great bought her entire collection.

What Louis XV wanted was not a model in wax at all. He wanted to be able to show the blood flowing through the body—not a dead anatomical prop, but a flexible, active automaton: an artificial, bleeding man. As it happens, this coincided perfectly with Vaucanson's own objectives. Until about 1730 he had been working on some automata he called "moving anatomies," intended for the purposes of medical teaching, but he had been forced to abandon them owing to lack of funds, and had turned instead to figures he knew would excite public interest. Now he was being asked by the King to return to his original project, but on a much more monumental scale. If his early androids had displeased the theologians, this one was his most God-tempting yet. He had made machines breathe, he had made them digest food, he had covered their hands in flesh, and now this "new Prometheus" was about to bestow on them the blood of life itself.

Louis had a reason for requesting this particular machine, and it was related to his support for the surgeons. Until the 1730s, surgeons had been regarded by the medical profession as mere craftsmen. They were represented by the guild of barber-surgeons rather than by any professional institution, and so were grouped more with barbers than with physicians. Surgeons, the physicians insisted, were unversed in philosophy or Latin and therefore did not have the intellectual means to establish a theory of their craft. They were just there to assist the physicians with their manual skills.

The surgeons rejected this view vociferously at the beginning of the eighteenth century, and the prime movers on that side of the debate were Mareschal, La Peyronie, and the latter's protégé François Quesnay. Their argument gained support in 1715 when Mareschal, first surgeon to Louis XIV, diagnosed the King's fatal illness early on, and was ignored by his chief physician, Fagon, who allowed the King to die. Fagon was dismissed, and the position of chief surgeon was never again seen as beneath that of the physician. Quesnay argued eloquently and often on the surgeons' behalf, pointing out that traditionally they had in fact been university educated, and not just apprentices to barbers. They had the *philosophes,* and more importantly the King, on their side, and in 1731 a Royal Academy of Surgeons was finally established.

Even then, the surgeons had to fight to be seen as equals, and the continuing dispute was conducted on the basis of certain medical issues, at the core of which was the question of bloodletting, a procedure performed by surgeons that was gaining theoretical ground. Though Harvey had discovered over a hundred years earlier that blood circulated in the body, little work had been done on how illnesses stemmed from troubled circulation, or how malformations of the system arose. "It is surprising," Quesnay wrote in 1730, "that since the discovery of the circulation of the blood, so few authors should have applied themselves to giving all the instructions and clarifications on this subject which can be drawn from the experiments and observations of the great masters."

The arguments over bloodletting were not only conducted in words. Experiments and clarifications came from practical replicas of the circulatory system. The first known model of the blood system was constructed by a German doctor and announced in the *Journal des Savants* in 1677. In his treatise on bloodletting Quesnay reported that he himself had built "a hydraulic machine" out of tin, which showed that the blood moved according to the laws of hydrostatics.

The head surgeon at the main hospital in Rouen, Claude-Nicolas Le Cat (whom Vaucanson had probably met as a student), was also at work on a machine by which he intended to prove his own theory of bloodletting. Le Cat was a brilliant, often argumentative surgeon and teacher of anatomy who for several years in a row won the annual prize awarded by the new Academy of Surgeons for the best essay on a given surgical subject. He described his machine as "an automated man in which the primary functions of animal economy, the circulation of the blood, respiration, and secretions can be seen to be executed, by means of which the mechanical effects of blood letting can be determined, and several interesting phenomena which do not appear to be susceptible to it can be subjected to experimentation."

In 1733 a young surgeon named Abraham Chovet exhibited in London what was described in the papers as a "new figure of Anatomy which represents a woman chained down upon a table, suppos'd opened alive; wherein the circulation of the blood is made visible through glass veins and arteries." Chovet pumped through the glass arteries and veins a red blood-like fluid, and claimed in the accompanying tract that "any Person, tho' unskilled in the knowledge of ANATOMY, may at one view be acquainted with the *Circulation of the Blood*, and in what Manner it is performed in our living Bodies."

So in asking the master mechanician to construct a model of the circulatory system, Louis was requesting not just a demonstration model but ammunition for a highly delicate debate. And the issues relating to blood were not confined to doctors and surgeons—they entered, to some extent, the popular imagination: there was even a machine exhibited at a Paris fair in 1736 that was said to "represent exactly the circulation of the blood."

Vaucanson started work straight away. In 1741, while in Lyon to inspect the silk-manufacturing process, he gave a lecture at that city's Academy of Art, of which Quesnay and Voltaire were associate members. The talk, which was highly unexpected, was reported in the Academy's minutes. The secretary noted:

> Monsieur Vaucanson, known for various automata which have received the approval of the Royal Academy of Sciences and the applause of the public in Paris, having come to this city, and our Director having given him permission to attend this session, told the Academy of a project he has imagined[:] that of constructing an automaton figure which will imitate in its movements animal functions, the circulation of the blood, respiration, digestion, the combination of muscles, tendons, nerves, etc. The author claims that by means of this automaton we will be able to conduct experiments on animal functions, from which we can make deductions in order to understand the different states of health of human beings and to heal their ills. This ingenious machine, which will represent a human body, may eventually be used for demonstration in an anatomy lesson.

Curiously, however, Vaucanson did not want his paper to be published afterwards. Towards the end of that year, another entry in the minutes explained that the author had not wanted to submit a copy for publication because, he said, "it is nothing but a project which may never be executed." It may have been modesty, or superstition, that prevented Vaucanson from making his idea fully public; but his biographers imply that there was a more calculated reason for his secrecy.

Vaucanson was to continue work on his artificial man for the next quarter of a century. He knew, of course, that what he had in mind would be difficult, if not impossible, to bring to fruition and, all the time he was engaged in more official pursuits, this artificial man was to be his private preoccupation. Meanwhile, there were many other attempts to build a blood machine, and Vaucanson appears to have seen no reason why he should give his rivals any clues about his own project. Another possibility is that, in keeping a low profile on this front, Vaucanson was able to quietly steal other people's ideas. Chovet's blood machine was in London, Quesnay was more of a theorist; his only serious threat was Le Cat.

Le Cat was happy to make himself unpopular in the service of higher aims. Although the Academy of Surgeons awarded him its gold medal several years in a row, he was never part of that coterie, and always refused to move to Paris from Rouen. Vaucanson was in the better social position by far; but Le Cat had been at work on his artificial man for longer and had more medical experience. He was instrumental in founding a separate Academy of Sciences in his own city (in 1744) and, after much wrangling, also succeeded in establishing a dissection theatre in his hospital, where he gave public anatomy classes that were extremely well attended. He was the inventor of a number of surgical instruments, and had a very public row with another surgeon, in which Le Cat supported more humane operative methods. He was made a nobleman four years before his death in 1768.

The year before Vaucanson gave his presentation in Lyon, word of his intentions had already spread, and Le Cat's supporters rallied round. In November 1740, Pierre-Robert Le Cornier de Cideville, a friend of Voltaire and future member of Rouen's Academy-to-be, wrote to Le Cat:

> You are working, so I am told, on your artificial man and you are right in doing so. You must not let Monsieur de Vaucanson accept the glory for ideas he may have borrowed from you. But he has applied himself only to mechanics, and has used all his shrewdness for that purpose—and he is not a man who is afraid to take extreme measures. I would advise you therefore to see, from practical experiment and from the actual construction of your machine, whether it will work in practice before you make your plans known to the public. So many machines strike us by their apparent possibility, only to be found wanting, in practice, in some area we had not worked out enough . . . Do not interrupt your work on this idea, which, if developed, will deserve to be presented to the Academy of Sciences in Paris, on behalf of our own nascent Academy, if in fact it is born.

Clearly, the lines of battle were drawn: Vaucanson had more experience in the manufacture of machines, and Le Cat knew more about the human body. Despite Le Cat's great prowess in the field of anatomy, he risked falling behind in the practical construction of his android, while Vaucanson, who was less worried about practicalities, took all the credit. Le Cat was not to be outdone. Fours years later, he presented his cybernetic project at the inaugural session of the Rouen Academy. Cideville reported the occasion in a private letter to the secretary of the Academy of Sciences in Paris. "[Le Cat's] automaton," he wrote, "will have breath, circulation, quasi-digestion, secretions, heart, lungs, liver and bladder, and, God forgive us, everything else. But it will become feverish, it will have to be bled and purged, and it will resemble a man too much."

God forgive us, it will resemble a man too much—this was exactly the problem with the artificial project: it frightened even the most respected thinkers. Le Cat may have meant his version as a demonstration model, but Vaucanson's ambitions seem to have swelled out of proportion—to dwell on it for so long, to be so calculating and secretive, to dream of creating artificial life in the most complete way becomes quite chilling, and is matched by Cideville's phrase, "He is not a man who is afraid to take extreme measures." As it happens, that comment was rather prescient.

Contrary to what Le Cat may have imagined, Vaucanson did have some practical concerns. He was not content to make his android out of glass or tin. He wanted to build a working, moving, artificial man and he knew that a different, more flexible material was required in order to simulate the circulatory

system properly. Until then, the ideal substance was not known ⌐
to exist. But something was about to be discovered on the other
side of the world.

In 1735 Louis had sent the scientist Charles Marie de la
Condamine on an expedition to South America, to measure a
degree of the meridian. He was accompanied by an astronomer, a doctor, a botanist, and a mechanician, and they
remained near the Equator for ten years. The *encyclopédiste*
Jean d'Alembert called the trip "the greatest enterprise ever
attempted by Science," and La Condamine's exploratory
methods proved so successful they became a model for subsequent scientific journeys, including those that attempted to
solve the problem of longitude. The travellers went through
hurricanes, earthquakes, and volcanic eruptions. When they
returned to Paris in 1745, they brought with them a number of
important discoveries. Some were geographical (La Condamine had drawn up a map of the province of Quito, in what
is now Ecuador), some were clinical (one of the Jussieu brothers, founders of the Museum of Natural History in Paris, came
back with cinchona, a bark that had great medicinal properties), others were anthropological or botanical.

Amongst their imported knowledge was the introduction
to Europe of a new material: the Amazonian Indians called it
"cahuchu," and Linnaeus later classed it as *Hevea Brasiliensis.*
Its "discovery" by the French was of course a piece of colonial
mythology—Indians in South and Central America had been
using it for centuries, and this had been noted by the Spanish
conquistadors. But the French were the first to bring any of it
back. La Condamine reported that the liquid substance could
be moulded into any shape, that once hardened it was resistant

to water, and that its most remarkable property was its great elasticity. It was just what Vaucanson had been waiting for. In other words: rubber.

During his early days in Paris Vaucanson had met Bertin, who was to become minister of finance, and was in charge of all commerce and industry. When Vaucanson heard of the explorers' discovery, he wrote to Bertin asking if some experiments could be conducted on the substance on his behalf. Bertin was also interested in the issues concerning the circulation of the blood, and knew of the King's support for Vaucanson's machine. He agreed to help him, but their inquiries had to preserve the utmost secrecy.

Although La Condamine had been introduced to rubber resin on his trip to Peru, he had not seen the tree it came from, and had little scientific knowledge of the plant. When Bertin came to him for advice, La Condamine referred him to another man, François Fresneau, who was posted in Cayenne, the capital of the French colony of Guyana. Bertin wrote a letter to Fresneau, including the most florid bribes ("I see that you are a good and faithful servant of the King," "I shall not forget to inform the King of your goodwill," etc.), and asking him specific questions about rubber without once mentioning Vaucanson. His initial inquiries were: firstly, if there was a way of resealing works made in the resin when they were pierced or otherwise damaged (for this, read attaching one tube to another, or an artery to a vein); secondly, if anyone in France had the resin, how the tree was cultivated, and whether it could be grown in France (implying that a large amount of the material was required); and lastly, if there was a liquid that could keep the resin moist but not penetrate or dissolve it (the blood content of the system, perhaps?).

The problem with the resin was that, while it was malleable in its liquid form, it could not be preserved in that form very long, and so, with research into its preservation or dissolution at such an early stage, it would be impossible to transport it to Paris in the quantities required. Fresneau had worked out a way of dissolving the dry rubber in walnut oil over a low heat, but when it reverted to its hardened state it lost all its elasticity, which clearly was of no use to Vaucanson. Fresneau presumed from the nature of Bertin's queries that he planned to make pipes, but he asked no more about his final objectives and set to work on finding the answers to his questions. Unfortunately, Fresneau concluded that the liquid resin was not transportable in large quantities, although spirit of terebenthine might produce the desired effect if the real spirit was available. However, in the course of his research he developed an indepth knowledge of the plant and, having sent his report to Bertin, asked La Condamine (with whom he had been corresponding) if a copy might be published in the proceedings of the Royal Academy of Sciences. La Condamine, who had been very supportive of Fresneau and had already presented a version of his early findings to the Academy, seemed well disposed to the idea; but it would be two years before La Condamine got hold of a copy of his report, and even then there was a problem. He wrote to Fresneau saying he would have to "excise all mention of the minister [Bertin], who would not think well of being written about . . . It is Monsieur de Vaucanson who made these requests, and they are for some

machine he is planning: I do not think he means large pipes."
So now Fresneau knew that Bertin's involvement was to be
kept secret, but it had been at the expense of revealing Vaucanson's. His report was returned to him with the words: "Do not
speak either of Monsieur Bertin or of the Court."

Fresneau rewrote his account, and re-sent it to Paris. It was
never published; and until the late nineteenth century the credit
for his research was given to La Condamine. What was the purpose of all this intrigue and deception? Was the project to be
kept secret because of Le Cat? Because of other potential competitors? Or simply because the scale of it—the exploration,
the botanical research, the chemical analysis and intercontinen-
tal correspondence—was so threateningly monumental? It's
hard to say, but the project did not stop there.

In 1763, the great French chemist Pierre-Joseph Macquer
received a package from Vaucanson. It was Fresneau's report,
and it was accompanied by a letter asking him to return it as
soon as he had read it. Macquer was brought in on the secret
project, and Fresneau's unpublished findings had become the
most crucial contraband. In fact, Macquer did discover that the
resin could be preserved in carefully rectified ether, which
could be allowed to evaporate without harming the rubber's
elasticity, but he only managed to collect it in small tubes of
about the thickness of a quill pen. The quantities were insuffi-
cient. At this point, the conspirators decided that if rubber
could not come to Paris, they would go to South America.

In his tribute to Vaucanson, Condorcet wrote that "an able
anatomist was sent to Guyana to preside over the work; the
King had approved the journey, he had even given orders for
it." It is not known how many people went to Guyana, or any-

thing more about the expedition. The project was eventually abandoned.

The lack of further information about the journey leaves a good deal of room for speculation. Why were all these people—the greatest scientists of the Enlightenment, the most powerful men in France, the ruler of the kingdom—so wrapped up in a plan to make life out of nothing? It's as if earthly pursuits were not enough; they had to test the limits of the Promethean dream.

The only reason given for the project's abandonment, after years of work, was that Vaucanson had become "disgusted." With what? With his own ambition? With the terrifying nature of his anatomical plans? It would appear that just as the fantasy, in which so much had been invested, was about to come true, the other side of it was seen; it became monstrous, and crumbled.

CHAPTER TWO

An Unreasonable Game

A thinking machine? *Was the Chess-Player such a creation? . . .*
A thinking machine! *Was such a thing really conceivable?*

—Joseph Friedrich Freiherr zu Racknitz, 1789

Could a machine think?—Could it be in pain?—Well, is the human body to be called such a machine? It surely comes as close as possible to being such a machine.

But a machine surely cannot think!—Is that an empirical statement? No. We only say of a human being and what is like one that it thinks. We also say it of dolls and no doubt of spirits too.

—Ludwig Wittgenstein, *Philosophical Investigations,* 1953

A few years after Vaucanson abandoned his project to make an artificial bleeding man, Wolfgang von Kempelen, a Hungarian man in his thirties, constructed an android that troubled him, even as it pleased his audiences. The Automaton Chess Player, built in 1769, was said to be "for the mind . . . what the Flute Player of M. de Vaucanson is for the ear." It was a machine that seemed to think, and it caused its inventor to be called, like Vaucanson before him, a "modern Prometheus." The Chess

Player was to elicit wonder throughout the world, but Kempelen, an eminent mechanician, insisted that it was only a toy, a trifle he had concocted for the amusement of Empress Maria Theresa. The machine's widespread popularity worried him; he dismantled it, inexplicably, soon after it was first shown.

The Chess Player was a carved-wood figure of a man in Turkish garb (white turban, striped silk shirt, fur-trimmed jacket, white-gloved hands, drooping moustache) sitting at a wooden chest, to which was affixed a large chessboard. The measurements given for the chest were between 4 and 4½ feet long, about 2 feet deep and between 3 and 3½ feet high. It had three doors in front (two of these opened on to a single compartment), and a single drawer at the bottom (which was decorated to look like two drawers). There were two small doors in the back of the cabinet, one of which was in the figure's spine. The player's right hand rested flat on the top, next to the chessboard. In his left hand, slightly raised from the surface, he held a long, thin pipe. His head moved on his neck, his eyes moved in their sockets. He had the appearance, as one visitor put it, "of someone who has just been smoking."

The figure, often known as "the Turk," was said to be about life-size, though some accounts say half that, and some say it was larger. The clothing was a reference to the part of the world where chess was thought to have originated, but its implications were not limited to the game. The outfit also had the psychological effect on audiences of a particular Orientalist fantasy: the unknown, the spirit-like forces of darkness, came dressed in the attire of the East.

Joseph II, Maria Theresa's successor, was engaged in an anti-Turkish diplomatic campaign when he asked Kempelen to reassemble the automaton. It had been twelve years since he had taken it to pieces, but Kempelen did as he was asked. The revived automaton was such an immediate success at court that Kempelen was granted a two-year leave of absence from his post as privy councillor, so that he and the Chess Player could make a tour of Europe.

This time, he gave in to his invention's popularity. Kempelen and the machine went from Vienna to Paris and Dresden, and in each city the automaton won every game (one of the Turk's merits was that it always—or almost always—won, even against the most illustrious opponents. It beat Napoleon on one occasion, and Catherine the Great was said to have been disqualified for cheating). In Paris it was presented to the Royal Academy of Sciences, where it played against François André Danican Philidor, the greatest chess player of the eighteenth century, who was especially known for playing blindfold.

Philidor's biographers report that Kempelen approached him before the game, and asked him to agree to lose against the automaton. Philidor apparently consented, with the proviso that the Turk must play well enough for him not to be ashamed to lose. In the end, it did not play well enough, and Philidor won the game, but he later told his eldest son that no other game of chess had ever left him so tired. It must have been like playing blindfold, only what was hidden was not the board, but the mysterious force behind his opponent.

It was suggested at other times that the machine was manned by some demon—there were reports that people crossed themselves on entering the room, and that ladies who attended the exhibition sometimes fainted from fear. What Kempelen called an "illusion," in the magician's or exhibitor's sense of the word, was often taken to be a different kind of illusion—a spectral one, a hallucination, or a communication with the dead.

In 1814, a year before he wrote "The Sandman," E. T. A. Hoffmann rehearsed his explorations of the Uncanny in another short story, entitled "The Automata." Hoffmann had read Wiegleb's *Instruction in Natural Magic,* which contains a long description of the Chess Player. He had attempted to follow the book's instructions for building automata, and visited androids whenever they were exhibited. In his story, a machine clearly based on Kempelen's is said to speak, and is attributed with psychic powers—it is asked questions, and in return it "seems to read the very depths of the questioner's soul." One of the characters comments that he finds distasteful "all figures of this sort, which can scarcely be said to counterfeit humanity so much as to travesty it—mere images of living death or inanimate life." The character imagines a scene that he thinks of as "fearful, unnatural, I may say terrible," and that was to become central to "The Sandman": a living man dancing, unwittingly, with a lifeless mechanical partner. "Could you look at such a sight, for an instant," he asks, "without horror?"

The fact was that Kempelen's Chess Player invited such reflections and fears. It dealt in the Uncanny, and trespassed on ground that was thought, insistently, to be the exclusive province of the living.

The automaton was exhibited in Savile Row in London in the years 1783–84. Aside from the admiring publicity it received, the show sparked a number of pamphlets, seeking to "expose and detect," as one of them claimed in its title, the nature of the mechanism. By 1790 over half a dozen of these had been published, and more were to come. It was a guessing game that became a polemic, and there was an urgency of argument and an anxiety in the writings about the chess automaton that say less about the machine than about the public it was shown to. The anxiety was connected with the question of what was human, and only human. A machine could not be intelligent, it was insisted. A machine could not play chess.

Though most people saw the Chess Player in an exhibition room somewhere on its tour of Europe, Karl Gottlieb von Windisch, a compatriot and personal friend of Kempelen's, wrote a pamphlet in which he described with great precision the place where the automaton was made and first seen. Kempelen lived with his wife, two sons, and two daughters in an apartment on the first floor of his house, and his workshop and study were on the second floor. Visitors would be received in the apartment before being led upstairs, where the automaton, which faced the door, was the first thing that met the eye. They had to cross the workshop in order to get to the study, and so would walk past a collection of carpenter's, blacksmith's, and clockmaker's tools, "thrown about in most admired confusion." The walls of the study, by contrast, were mostly covered by large oak dressers with heavy double doors at the base and glass at the top. Some contained books, others antiques, and the ones at the far end of the room housed a small natural history collection. The empty spaces between the dressers were decorated with prints and drawings done by Kempelen himself. The automaton was in the centre of the room.

Wherever it was shown, the manner of the exhibition remained the same, and involved an elaborate display ritual, much examined by would-be unveilers. Kempelen would announce to the assembled audience that he was about to show them the mechanism. He would open the door on the left (the viewer's left) to reveal a compartment lined with dark-coloured cloth and filled with cogs, cylinders, wheels—which looked like the insides of a clock and a barrel-organ combined.

The exhibitor would then move to the back of the cabinet and, opening the door directly behind the first, shine a candle through the opening, so that the absence of anything but machinery was fully visible. He then locked the door at the back. Next to be opened was the drawer at the bottom, from which would be removed chess pieces and a cushion for the fig-ure's elbow. The drawer would remain open as the exhibitor revealed the interior of the main compartment by opening the two doors on the viewer's right. This space was lined, like the adjoining one, with dark fabric. It was mostly empty, except for two quadrant-shaped pieces of metal near the top, and what looked like wires connecting them. At this point, all the doors and the drawer at the front were (or appeared to be) open.

The exhibitor would turn the Chess Player around so that the back was visible to the audience, and lift the back of the fig-ure's jacket to reveal two more compartments, one in what might be called the figure's spine, containing a cylindrical mechanism not unlike a piano pedal, and a smaller one in the thigh, with a line of wire running through it. The audience could now see that the figure was seated on a decoratively carved wooden stool, which was attached to the cabinet. Bunches of cloth made up Turkish pantaloons, and the feet were crossed in a feigned side-saddle position.

The idea was to make the audience think they were seeing the whole interior at once or, as Windisch put it, seeing "the

Automaton stripped naked." His humanesque phrase is apt. In Windisch's book, in which Kempelen's original engravings are reproduced, the view from behind is the most unsettling. From the front one can clearly see a face, or a simulacrum of a face, but here the Player hardly looks like a man. The fur-trimmed jacket is pushed back over the shoulders, making a lumpen image of the human form, and it is hard to tell at first what the engraving represents. The back of the head is the back of the turban, which is wrapped around a kind of fez, and has a knot tied in the centre. The shrivelled, pushed-back fabric legs (unseen when one looks directly at the upright torso) look as though they might have suffered, as a machine couldn't, from polio. The eye works back into the picture from the feet (simulated slippers), and once the mannish form is made out, there is something terrible about the boxed-in levers that replace the spine, and something shameful about looking at it, behind its back, so to speak, as if one were catching someone's private bionics unawares.

Once the audience had seen the Player from all sides, the exhibitor would wheel it back to its original position. He closed all the doors and the drawer. He occupied himself for a few moments at the back, as if adjusting the mechanism. This was the only time he touched the automaton, except to wind it up with a key, which he did next, and subsequently after every ten or twelve moves. An opponent was found from the audience and seated at a separate table, with his or her own chess men (some accounts say this was a later development, and that at first the opponent played with the automaton's pieces). When its opponent made a move, the automaton would survey the board with his head, sometimes roll his eyes, before "deciding" how to play. With each of its moves, a muffled sound of wheels was heard, like that of a grandfather clock. Check would be indicated by two nods of the head, checkmate by three. If the opponent made a false move (as many did, trying to trick the automaton), the Chess Player would shake its head and return the piece to its previous position.

The theories that addressed the mystery fell, broadly, into two categories, and each pamphleteer generally attempted an explanation using each one in turn. The first possibility was that the machine was operated by Kempelen using some form of remote control—strings, wires, magnets. But this was explained away by the fact that magnets and loadstones were permitted in the room; and Windisch, in his description of Kempelen's studio, intended to dispose of the possibility of a connection with an adjoining apartment. The second theory was that there was a person inside the cabinet. Most of the guessers' attention was devoted to this: if there was someone inside, where could they be when the interior was displayed? How big could they be? How could they see the game? By what mechanism could they play?

In 1784, when the Chess Player was on show in London, Philip Thicknesse, a writer credited with the "discovery" of Thomas Gainsborough and described by his biographer as "the most quarrelsome man who ever lived," wrote a pamphlet disputing the claims of the automaton. That this machine could be made to move chess pieces "as a *sagacious Player*," he wrote, was "UTTERLY IMPOSSIBLE: And therefore to call it AN AUTOMATON, is an imposition, and merits a public detection." Thicknesse described an added trick performed by Kempelen, which would have the audience believe that it was the inventor himself who had "incomprehensible and invisible powers":

> He always places himself close to the right elbow of the Automaton, previous to its move; then puts his left hand into this coat pocket, and by an awkward kind of motion, induces most people to believe, that he has a Magnet concealed in his pocket, by which he can direct the movement of the Turk's arm at pleasure. Add to this, that he has a little cabinet on a side-table, which he now and then unlocks, and locks; a candle burning; and a key to *wind up* the AUTOMATON; all of which are merely to puzzle the spectators: for he takes care that they shall see him move his hand and fingers in his pocket backwards and forwards, on purpose to enforce the suspicion that *he, not an invisible Agent,* is the antagonist against whom you play; whereas, he is only a party in the deception; and the real mover is concealed in the Counter.

It's an odd distraction, a double decoy: Kempelen diverts attention from one solution, the inside-agent theory, by directing the audience towards another, the remote-control theory—neither of which furthers the Chess Player's claims to automation. Kempelen, it was often forgotten, never said his invention was a bona fide automaton. He spoke of it variously as an "illusion," a "trifle," a "*bagatelle,*" as Windisch reports, "which was not without merit in point of mechanism, but the effects of it appeared so marvellous only from the boldness of the conception, and the fortunate choice of the methods adopted for promoting the illusion." The decoy of fumbling with a magnet in his pocket would be in keeping with this, as would the fact that the cabinet itself was done up in seemingly unnecessary trompe l'oeil—two doors opening on to one compartment, one drawer made to look like two. Illusion was the governing property of the Automaton Chess Player: it was part of its style, and made up its world; there was more than enough illusion to spare.

What Thicknesse called the "whole delusion" was supported, he claimed, by "an invisible confederate." The automaton was only exhibited for an hour at a time—from one until two in the afternoon, when he saw it (because, as he thought, the person inside could not stand a longer confinement). There was enough space, he estimated, exclusive of clockwork, to contain "a child of ten, twelve, or fourteen years of age; and I have children who could play well at chess, at those ages." All the player had to do in order to follow the game was to look at a mirror fixed to the ceiling of the cabinet, and to practise reading the game backwards. Later in the pamphlet, having seen a sleeve move when it shouldn't have done, Thicknesse decides the invisible player must be able to see everything through the hair trimmings in the Turk's jacket and perhaps has no need of a mirror. Whatever the method, his conclusion about the internal operator remains the same, and Thicknesse ends with a wonderful piece of throwaway sarcasm. Forty years earlier, he says, he saw 300 people, who had each paid a shilling, gathered round what purported to be "a coach which went without horses." He continues:

The Duke of Athol, and many persons present, were angry with me for saying it was trod round by a man within the hoop, or hinder wheels; but a small paper of snuff, put into the wheel, soon convinced every person present, that it could not only move, but sneeze too, *perfectly like a Christian*. That machine was not a wheel within a wheel, but a Man within a wheel . . . and the Automaton Chess Player is *a man within a man;* for whatever his outward form be composed of, he bears a living soul within.

If that was the case, whose "living soul" was it? The year after Thicknesse's theories were published, an anonymous pamphlet appeared in Paris, claiming that the automaton concealed a dwarf who was a famous chess player (though the identity of the dwarf was not divulged). His legs were concealed, according to this pamphlet, in two hollow cylinders, with the rest of the body outside the box, hidden by the "petticoats" of the Turk. When Kempelen wound up the machine, it was only to make enough noise to cover up the sound of the dwarf returning to the cabinet, and as the object was slowly wheeled about the room, the traps by which he got there were shut.

Joseph Friedrich Freiherr zu Racknitz saw the machine in Dresden and in 1789 wrote a pamphlet on it. He didn't specify who he thought was inside, but suggested that his ideas coincided with those of Henri Decremps, who subscribed to the dwarf theory. Racknitz was often misrepresented, however, since his findings were propagated by one Thomas Collinson, who met him in 1790 and understood him to have thought the machine was inhabited, as Collinson wrote in his Supplement to Thomas Hutton's *Philosophical and Mathematical Dictionary*, by a "well-taught boy, very thin and tall of his age."

Racknitz suspected that the drawer at the base of the cabinet did not reach all the way to the back, and thought that a person could lie in the space behind the drawer while the exhibitor showed the interior of the cabinet. To the question of how the player would see the game, he answered that each chess piece had a magnet on the bottom, and each square a pointer on the underside of the board. The pointer moved when a piece was put on that square. The player, Racknitz believed, had his own, smaller chessboard inside, numbered to match the chessboard on the ceiling. He could make the Turk's arm move using a pantographic device that would duplicate the move he made on his own board.

Racknitz went to more trouble than the others to imagine the conditions of the hidden player. His thoughts were not only of the impossibility of a machine having human faculties, but also of the actual needs of human beings. If there was a person in the box, he wondered, how would they see in the dark? There would have to be room for candles. How would they breathe? There would have to be holes in the bottom. And so, gradually, Racknitz constructed an imaginary domain for his imagined player, and the illustration he gives looks as impossible as a surrealist collage. In Racknitz's engravings the child or dwarf is a perfect miniature man (is this how Racknitz imagined a prodigy to look?). He has made himself improbably at home—two candles burn (as on a mantelpiece) on a ledge where his arm can rest, and on which he places the pieces taken. He is much smaller than the Turk, who seems to reign over him, as if the living man were the puppet, and not the other way around.

Despite the seeming improbability of the drawing, Kempelen appeared to be affected by Racknitz's theories, and announced that he was going to take the automaton apart again. At the very least, Racknitz had correctly imagined the pantographic device, and it was the part of the Chess Player of which Kempelen was most proud. Indeed, a close look at his friend Windisch's pamphlet shows that he had already given that part of the game away: "The invention of a mechanical arm," it reads, "whose movements are so natural, which grasps, lifts and sets down all with such grace, even if this arm were directed by the two hands of the inventor himself, it alone is so complicated that it would ensure the reputation of many an artist."

Racknitz's drawings of the arm, which a later commentator said were so close to Kempelen's design as to raise suspicions that the secret had been disclosed to him, show the workings of a skeletal marionette. The design of the arm was an important piece of engineering—and in an inversion characteristic of the issues raised by the chess automaton, the device contributed to subsequent advances in the development of artificial limbs. The automaton helped to design the human. Kempelen died in 1804, fifteen years after Racknitz's book was published, with the disassembled automaton still in his possession.

After Kempelen's death his son sold the Chess Player to Johann Nepomuk Maelzel, a man often credited with the invention of the metronome. Maelzel, the son of an organ-builder, was trained as a musician and was at the time of Kempelen's death court mechanician or, as one evocative translation put it, "philosophical instrument maker," to the Hapsburgs. He was a close friend of Beethoven, whom he persuaded to compose what became his "Battle" Symphony (Opus 91), for Maelzel's Panharmonicon, an automated orchestra of forty-two mechanical musicians.

In 1818, Maelzel and the machine embarked on the Chess Player's second European tour. It was exhibited in Paris, and in London was shown at a small room in Spring Gardens, next door to a popular exhibition of automata. Maelzel's exhibition included some other inventions and possessions of his own. There were, at various stages of the tour, some rope dancers, an automaton trumpeter, a mechanical instrument called the Orchestrion, which imitated a full military band, and a moving panorama called the Conflagration of Moscow. Maelzel added some improvements to the Chess Player—rather than nodding its head, the Turk could now say "*échec*" (check) by means of a set of bellows.

The guessing game started up again, with the same theories put forward. A writer calling himself an "Oxford Graduate," who like his predecessors considered it "absolutely impossible that any piece of mechanism should be invented which . . . should appear to exert the intelligence of a reasoning agent," proposed that it was operated from the outside by means of a hidden wire. Unfazed, Maelzel toured the country and took the automaton to Scotland, before returning to a larger, better-fitted exhibition room in St. James's Street in 1820. After the Oxford Graduate came a Cambridge undergraduate, Robert Willis. Willis was the son of mad King George's doctor, and he went on to become a distinguished mathematician and professor of mechanics. Willis took his cue from the excess of illusion; the machinery, he wrote, was all a hoax.

He smuggled an umbrella into the exhibition (he had been to the earlier venue, where the space was conveniently cramped) in order to ascertain the exact measurements of the cabinet. He deduced from several visits that the space inside was larger than it seemed and that a full-grown man could position himself in such a way as to be unseen when each part of the interior was shown—the effect of seeing the entire thing at once was just that: an effect. Willis argued that a man shifted between the two portions of the cabinet as the doors were opened, and that when they were closed he climbed into the figure so that his head was level with the Turk's chest. He saw the game through the figure's diaphanous clothes and, Willis suggested, put his arm into the figure's sleeve.

Of all those who wrote on the Chess Player, Willis was perhaps the most severe in his insistence that no machine could make decisions governed by reason. "It cannot," he wrote, "be made to vary its operations so as to meet the ever-varying circumstances of a game of chess. This is the province of the intellect alone." "It cannot," he concluded, "usurp and exercise the faculties of the human mind."

Willis states perfectly the anxieties floating beneath the surface of his logic. To say that the machine could not exercise the faculties of the human mind would have been enough; but in order to exercise those faculties, according to Willis, the machine must first "usurp" them, knock them out of the human domain where they rightfully belonged. Exercise then

12/12/2007

becomes not simply a question of possibility or practicality, but a question of ethics. The bottom-line fear Willis gives voice to in that one worried word—"usurp"—is that there may be nothing that separates the powers of machines and men.

Maelzel was undeterred by these writings, and in 1826 he went to America. A couple of years later in Baltimore two boys, who had been looking down from a rooftop opposite Maelzel's exhibition room, claimed to have seen a person crawling out of the automaton after a show. The story was printed in the local paper: "The ingenious contrivance of Mr. Kempelen," ran an article in the *Baltimore Gazette*, ". . . after a period of nearly sixty years of doubt as to the mode of its operation, has at length been discovered by accident to be merely the case in which a human agent has always been concealed when exhibited to an audience." But because the story provided such good publicity, a Washington newspaper spread the rumour that Maelzel himself had manufactured it expressly for that purpose.

Maelzel travelled around America for over ten years, and in Boston he met Phineas Taylor Barnum, who was just starting out with his first exhibit, Joice Heth, "the 161 year old woman," who was said to have been George Washington's nurse. Humbug, as he called it, was Barnum's business: Joice Heth was in her eighties. The question of whether she was truly as old as was claimed was part of what kept audiences coming. By the time Barnum and Heth arrived in Boston, however, ticket sales were tapering; meeting Maelzel gave Barnum an idea. He wrote an anonymous letter to a newspaper claiming that Joice Heth was a fake, that she was really a "curiously constructed automaton, made up of whalebone, India-rubber, and numberless springs." The crowds flooded back to see how they had been duped.

Maelzel and Barnum were quite different figures. Maelzel did not just attract audiences on the basis of his strengths as an exhibitor—though he was, so his contemporaries said, a perfect gentleman, and loved children, often reserving the front row for them. He was popular primarily because his reputation as a distinguished inventor preceded him. Barnum's game was showmanship and hokum, and he was the first to turn it into an art form that appealed to the masses. But their encounter helps to demonstrate how exhibitions of curiosities inhabited more than one world—collections of natural history, inventions of scientific ingenuity, human wonders, and feats of magic were often housed together (as in Kempelen's study) or practised side by side.

Maelzel told Barnum he would go far, because he understood the power of the press. Only a few months later the press wielded its power against Maelzel when the *Southern Literary Messenger* published Edgar Allan Poe's article on the Chess Player. Cast as an exposé, it generated a good deal of interest in the automaton as a hoax, though in terms of unveiling Poe was rather behind, since more, as we shall see, was already known.

Poe compared the chess automaton with another contemporary exhibit—Charles Babbage's calculating machine, which had been shown at the Adelaide Gallery in London the previous year. "What shall we think," wrote Poe,

> of an engine of wood and metal which can not only
> compute astronomical and navigation tables to any

given extent, but render the exactitude of its operations mathematically certain through its power of correcting its possible errors? What shall we think of a machine which can not only accomplish all this, but actually print off its elaborate results, when obtained, without the slightest intervention of the intellect of man? It will, perhaps, be said in reply, that a machine as we have described is altogether above comparison with the Chess Player of Maelzel. By no means—it is altogether beneath it—that is to say, provided we assume (what should never for a moment be assumed) that the Chess Player is a *pure machine,* and performs its operations without any immediate human agency.

Poe goes on teasingly to discuss the virtue of a thing he believes does not exist—a machine that can, on its own, make successive moves which are not "fixed and determinate," as they are in mathematics. Until he repeats Willis's theory, the entire essay is devoted to the elaboration of a fantasy hypothesis; but then, what the automaton caused its audience to imagine was part of its strength and its appeal: later in the nineteenth century, the famous magician Robert-Houdin reported a far-fetched version of the automaton's history, involving a Polish soldier whose legs had been amputated, and for whom the Chess Player was said to have been constructed as a kind of Trojan horse, in order to smuggle him across the Russian border. This story became a play, and then a silent film: the myth developed a life of its own, parallel to the historical facts.

Poe raises the question of whether or not the Chess Player is realistic, of how "human" it really is. As an imitation of life, he says, the invention is "indifferent." "The countenance evinces no ingenuity," Poe writes, "and is surpassed, in its resemblance to the human face, by the commonest of waxworks. The eyes roll unnaturally in the head, without any corresponding notions of the lids or brows. The arm, particularly, performs its operations in an exceedingly stiff, awkward, jerking, and rectangular manner." But these defects, he concludes, must be intentional, since Maelzel is clearly able to do a more life-like job. His rope dancers, according to Poe, are "so free from the semblance of artificiality, that, were it not for the diminutiveness of their size . . . it would be difficult to convince any assemblage of persons that these wooden automata were not living creatures." (The rope dancers were not in fact invented by Maelzel, although he almost certainly made improvements to them. He had bought them from a peasant in the Black Forest.) So Kempelen, and Maelzel after him, Poe supposes, must have made the Chess Player more machine-like than life-like, expressly to avoid the supposition that its movements were in fact guided by a human being.

There was another point in the Turk where the opposition between life and imitation became an issue. He played left-handed. Why would a machine, designed and constructed along lines of perfection, have the quirk or, some would say, failing, of a human being? This business was so bothersome that many believed it to be the solution to the mystery. Racknitz thought the arm was designed to make it easier for the internal player to see the game in a mirror on the ceiling. Windisch, having clearly been confronted on the matter, tried to wave it away by saying that he had it on the assurance of the inventor that he hadn't realized he'd made it left-handed until it was too late; and anyway, says Windisch, it didn't matter to Kempelen, so why should it matter to us? Do we care, he writes airily, if Titian painted with his right or his left hand? Intentionally or not, Windisch gives us the wonderful image of a mistake creeping into Kempelen's creation, amidst all the precision of clockwork and costumes: it's a picture of science, or mechanism, sharing a world with accident.

Poe uses the Turk's left-handedness as the last point in his essay, concluding it, as he thinks, with the indisputable argument that the left-handed play proves there must be a man inside. Poe's discussion rests on the assumption that all human chess players are right-handed—so there is, at the outset, a sort of rigidity of opinion on the matter of what is human; in a rational argument of this kind there is no room for the very things that divide humans from machines—preference, individuality, variety of ability. The Chess Player, Poe writes, "plays precisely as a man *would not*." His ideas about right- and left-handedness are "sufficient of themselves," he thinks, "to suggest the notion of a man in the interior." The operator needs to play with his right hand, and because of the construction that Poe assumes to be inside the box, he can only operate the arm of the automaton with his right hand if the arm is the automaton's left. But since Poe had no reliable information about the interior machinery, we have only his detective-like deductions to go on, and those break down as: it must *not* be a machine—that is, it must in reality be a man—precisely because it plays as a man *would not*. This is one of the essays in which Poe is said to have developed his famous powers of rational deduction, but the logic here is nothing but tangled—an indication of the reason-defying, or reason-threatening, nature of the problem.

The automaton ended its tour when Maelzel died of yellow fever on board a ship back to the United States from Havana. The boat belonged to a Mr. Ohl of Philadelphia, whose storeroom Maelzel had often used over the previous few years. The automaton was found there, divided up and stored in five separate boxes, with bits of other machines thrown in, so that it would be impossible for anyone but the exhibitor to reconstruct it. Maelzel owed Mr. Ohl some money when he died, and Ohl recouped it by selling off all of Maelzel's effects, in September 1838. At the auction, he bought the automaton himself, in all its boxes, but couldn't do anything with it; and after a year had passed, in which the boxes had been left untouched, he sold it on to a local doctor and professor of medicine, John Kearsley Mitchell. Mitchell was also a naturalist, and had made some mechanical objects himself. He spread the cost of the automaton amongst a group of subscribers, so that the object was owned by a club.

He then proceeded to reconstruct it, digging about in still more boxes for parts that he thought must be missing. The *Philadelphia Gazette* reported at the time that "such a chaos of moving eyes, wooden membranes and brass muscles, the Doctor had probably never encountered at his dissecting room; but his energy and perseverance surmounted all difficulties, and he had the satisfaction of recreating, and to give the requisite animation to the original form of the Turk." It's a lovely image—a surgeon piecing together, and resurrecting, a humanoid monster.

An Unreasonable Game

The reconstructed automaton had a brief third life, shown amongst the subscribers and their friends, until they chose to deposit it, for permanent but for the most part motionless exhibition, in what was known as the Chinese Museum, or Peale's Museum, in Philadelphia (someone later recalled having seen "the *dead* automaton at Peale's museum," meaning that it no longer performed). The building had once housed the great cabinet of curiosities of Charles Wilson Peale, painter and collector of marvels of all kinds. It seems appropriate that the Chess Player should have met its end at the home of this wonderworker, just at the time when Peale's own collection, which was opened as a museum in Kempelen's era, was being bought up and scattered by none other than Phineas Taylor Barnum. In 1854, a fire broke out in the National Theatre, a short distance away, and spread to the museum. There would have been time to save the automaton, a local man later pointed out, but by then everyone had forgotten it existed. The Chess Player was destroyed for good.

The secret, however, did not die with it. In 1834, an article appeared in a French penny magazine, the *Magazine pittoresque*, which revealed that the Turk had always hidden a man. Willis, it turned out, had been closest to guessing how the person was hidden—by slipping into each compartment as the other was shown (hiding in the dark where the candlelight wasn't being shone through), but he hadn't given Kempelen credit for the mechanical arm. Racknitz, who had on the other hand failed to guess where the person was hidden, got the arm, the magnets in the chess pieces, and the smaller chessboard inside, exactly right—what had looked in his engraving like an improbable collage was the impossible-seeming truth: a man sitting scrunched inside the box for an hour, sometimes more, his travelling chessboard illuminated by a candle, his brain occupied in the concentrated intellectual pursuit of a game of chess.

Why had it taken so long for the secret to become known? The Automaton Chess Player fulfilled what the historian Richard Altick has called "the public's desire to be baffled." It didn't matter how many times its inventor insisted the automaton was merely an "illusion"; it was constructed during the Age of Reason, yet many found reason less appealing than enchantment. Henri Decremps, author of an eighteenth-century handbook on magic, wrote of "the tint of charlatanism so necessary in this century if one is to obtain the vote of the masses." False automata met this need even more than real ones. Vaucanson's automata had set a new precedent: their inner workings were revealed as part of the performance, and this process was designed to sift out the charlatans. But it only gave trickery a new idiom: subsequent shows of false automata involved not only apparent mechanization, but apparent revelation as well. Pretending to unveil the inside of the machine was all part of the act.

It didn't matter if the Chess Player was an authentic automaton; it was, as its history shows so clearly, less an admirable piece of mechanism than a philosophical game. Audiences could be titillated by the possibility of automation; they could, to their mind's content, tempt fate and fear with the idea that machines could be like humans, without ever having to deal with the reality. It was like playing with machinery, or playing with what was human, the way one might play with fire. The label "a new Prometheus" was both an honour and a warning, since the truly Promethean territory was this: it was not mechanical ingenuity, the giving of imitated life, that had earned Kempelen his moniker, but rather the act of playing with life, and the dangerous thrill of the riddle his invention proposed.

The secret had been sold to the *Magazine pittoresque* by one of the automaton's directors (as the men inside were called). Once it was out, it was retold in *Le Palamède*, the first chess periodical, and a magazine of such distinction that news of the story's veracity spread all over the world. Several American newspapers reprinted the *Palamède* article, and a copy of it, in the *National Gazette* of 6 February 1837, was found amongst Maelzel's possessions when he died.

Inside a small, square leather book are pages made up to look like miniature chessboards. Some of them are empty, others have little black and red figures carefully pasted in, in varying positions. In heavy black ink, at the top and bottom of each filled page, is a number—the number of pieces remaining on each side. On the cover, etched in gold into the gleaming green leather, are the letters "J. MAELZEL."

I am in the Library Company of Philadelphia, taking objects and papers from a box full of Maelzel's things. There aren't many, but they are what he left behind in this city when he died: this chess problem book, with the endgames played by the automaton mapped out inside; some newspaper clippings; an anonymous letter sent to him in New York, threatening to expose the automaton as a fake; a wooden chessboard with holes in it, used for travelling—the sort the director of the automaton used inside the box, and probably the one used by

Maelzel on his last journey back from Havana, when he sat on the deck and invited the ship's captain to play against him.

Benjamin Franklin set up the Library Company in 1731 as a subscription library for his friends. Franklin was one of the earliest-known chess players in America; he wrote an essay on the subject, "The Morals of Chess," and when in France had frequented the Café de la Régence, in Paris, where all the great players were to be found at their game. He knew Kempelen, and had played against the automaton. True to his legacy, the Library Company now has amongst its rare books and manuscripts the collection of a nineteenth-century Philadelphian named George Allen. Allen, who died in 1876, wrote a biography of Philidor, and assembled a collection of books on chess that are like prized jewels to those who seek them. When the collection was sold two years after Allen's death, his executors compiled a catalogue and, in tones that sometimes read like a secret code, highlighted some of the obscure delicacies within. Last, and certainly not least, of these precious papers is, in the executors' description, "a very valuable private Autograph Letter from Mr. Lewis to Dr. Allen giving the history of his connection with the Automaton during Maelzel's sojourn in London."

William Lewis, one of the greatest chess players of the nineteenth century and the author of a number of treatises on the game, operated the automaton in 1818 and 1819. The manuscript memoir and letters he wrote to Allen are priceless, and the main reason I have come here; but Lewis was not the automaton's only director, and for anyone wanting to find out about these hidden inhabitants, Allen's collection is a repository of scattered clues.

An Unreasonable Game

The world of Allen's correspondents is far from that of the philosophical detectives who wrote the tracts on the Automaton Chess Player. Those people were thinkers, guessers, writers. These are chess players, a special coterie, like a magic circle, spread out over a number of countries and states. Their question, at the time of the automaton's exhibition, was not *whether* there was a person inside it, but *who*—which of their chess-playing colleagues—was inside it. More people knew about the existence of a director than could possibly be imagined from reading the published pamphlets. On the subject of a famous chess player who turned out to have been in on the secret, one cognoscente writes to Allen:

> You seem surprised at what I said about Deschapelles, but it must be obvious to you that above all others, Maelzel had to take Deschapelles into his confidence, for he was then the strongest player in France. In fact Maelzel had to conciliate *all* the strong players, for fear of losing games. And it is remarkable that the secret was so well preserved, considering the great number that were made acquainted with it. This circumstance is certainly honourable to Chess players.

Indeed, Maelzel hired many of his directors from the same virtuoso pool—amongst the players at the Café de la Régence, a sort of chess Mecca, or at the first chess club there, the Hôtel de l'Echiquier, run by Aaron Alexandre. The job of the automaton's director, far from being unknown to these experts, was one that many were keen to try. William Lewis, in describing how Maelzel approached him to direct the automaton, shows not the slightest shock on hearing that the exhibitor needed to hire someone for that purpose: "He wished to know if I was disposed and could find time to direct the moves," Lewis writes in his manuscript memoir. "I informed him that I would willingly do so, but could not afford much time." The moment of revelation is briefly dealt with: "Having agreed to the terms proposed & which were not too liberal I visited the Automaton and was initiated into its mysteries and after a little practise [sic] was well able to play my part."

Though Allen addressed Lewis as the sole surviving director of the automaton, preserved in his files are first-person accounts of being inside the Turk, not only from Lewis, but from two more, one of whom specifically asked Allen not to mention him in the essay he eventually wrote on the automaton's American tour; and there is information about others, since dead but once known to these people. Some of them operated it for a season or more, some were just stand-ins or decoys; a few were employed at the last minute and only played the endgames already laid out in the green leather book; many were the best chess players of their time. Their names emerge, like one jack-in-the-box after another.

None of those who directed it in Kempelen's time is known, but it was Johann Baptist Allgaier, a German chess player and author, who played against Napoleon at Schoenn-brunn (where Kempelen had designed the hydraulic fountains) in 1809 and won. At one time, it was said to have been directed by a young girl. Her name has been forgotten.

Aaron Alexandre, a rabbi, teacher, and the owner of the Hôtel de l'Echiquier chess club, made children's toys and mechanical objects in his spare time; he operated the automaton for a while in 1818 but couldn't bear the confined space and stopped. (Alexandre owned the building that housed l'Echiquier, and his pupils were lodgers; but after his wife's death he forgot to collect his debts and went bankrupt—when the building was sold an old chess teacher had lived there free of charge for ten years.) There was Weyle, nicknamed *"le petit juif,"* and Boncourt, an habitué of the Café de la Régence who left his mark on the automaton's history by sneezing during a game and causing Maelzel to alter the mechanism in case it happened again.

William Lewis was a merchant from Birmingham who was living in Soho with his mother when Maelzel approached him. Lewis operated the Turk during its first season in London, in the evenings and on Saturdays, when it played full games (in that time the automaton never lost a single game). The rest of the time the automaton only played problems, and on those occasions it was inhabited by a Mr. Hunneman, who was not as good a player but was able to direct it for these limited events. (Later, in 1820, Hunneman took down and published fifty games played by the automaton.)

Lewis tells a wonderful story in his memoir of the automaton, which gives a good sense of the guessing games of the chess coterie:

> Conjectures soon began to be made as to the person who directed the Automaton. At last I became suspected, and it seemed to Maelzel desirable to mystify the Public on this point. Accordingly I composed a Muzio Gambit and asked a friend of mine, a German on whom I could depend, to learn the moves and play

them against the Aut. (directed by Mr. H[unneman])
on a given evening when I knew I should be engaged
at a party to meet several chess players of the Club.
The scheme was quite successful; the game was a brilliant one and being known the next day at the Club by
those who were present with me at the party, the sus-
picion about me died away.

Lewis explains that it was extremely important to Maelzel that
these suspicions be quashed, since he made his money from
people who kept coming back to the exhibition, paying their
half a crown each, either to try their luck against the machine
again, or because they had heard that a good player was going
to challenge it and wanted to watch. The scheme worked, as
Lewis shows: "An old & valued friend of mine who played frequently told me he was sure he could beat the Aut. though he
somehow or other always lost: when he afterwards learned that
I was the Director he said, if he had known that he should have
saved many half crowns."

When Maelzel wanted to tour the country after the London season, Lewis's work commitments prevented him from
accompanying him, and Maelzel took with him a French player
named Jacques-François Mouret, the first to give pawn and
move to his opponents, and the person inside the box when the
games Hunneman published were played.

Maelzel left for America having primed a player in France
to meet him there later, when he could afford to pay for his trip.
In the meantime, Maelzel's exhibition opened at the National
Hotel on Broadway in New York, hiding inside the box a

woman—the wife of the man who operated the rope dancers. She was, wrote one informant, "in no way remarkable or distinguished for anything, her name I have forgotten." George Allen, in his essay, is very disparaging about her abilities, mainly, it seems, on account of her sex: he thinks it likely that it took Maelzel so long to set up his exhibition in New York (two months) because he had to train this director in "the management of the mechanism and in the playing of his select endgames." "This will be better understood," Allen goes on, "when I say that Maelzel was obliged, at the outset, to confide the direction of his Automaton to a *woman* [the italics are his] . . . Such an extemporized *Directress* must, of course, have required a good deal of instruction."

Once, when someone noticed that a Frenchwoman could be seen at all times except when the automaton was exhibited, Maelzel engaged as a decoy a young man named Coleman, the son of the editor of the *New York Evening Post,* whom he had befriended on his arrival. The Frenchwoman then played the part of the opponent, and suspicions were allayed. When she and her husband deserted Maelzel that season in New York, he hired Professor Anderson, a scientist and keen chess player who taught at Columbia University.

A young German man, who regularly visited the automaton's exhibitions in New York, introduced himself to Maelzel on one occasion. He was studying medicine, and amongst his successful experiments in what he called "natural philosophy" was the construction of several powerful magnets. The man (or boy—he was only sixteen) was called Charles F. Schmidt, and it was he who wrote to Allen requesting that his name not be disclosed in connection with the Turk. Though Allen could not use much of them, his letters are in many ways the most informative and eloquent in Allen's files.

Schmidt happened to mention his magnets to Maelzel one day, and Maelzel asked him to do him a favour, in return for which he would show him the secret of the automaton. In fact, the favour was not unrelated to the secret: the magnets in the chess pieces were beginning to lose some of their power, and Maelzel was stuck; he had no way of recharging them without destroying his livelihood, and no way to make the automaton work without the magnets. Schmidt's casual mention was therefore rather timely; he charged the whole set again, and waited for Maelzel to fulfil his part of the bargain. Later that summer, Schmidt and Maelzel found themselves alone with the automaton one afternoon. "You may imagine my astonishment," Schmidt wrote to Allen, "when . . . he asked me to assist him to lift off the cover of the Automaton, when out stepped Prof. Anderson; who by the by was quite tall. I was of course all eagerness to go in myself."

Schmidt did go in, and remained with Maelzel, travelling to Boston and directing the automaton there for a period of six weeks. "Though never a *strong* player," he writes of his own abilities, "there was at that time none better than myself."

Eventually, Maelzel had enough money to send for Wilhelm Schlumberger, the automaton's last director, who arrived from Paris in September 1826. Schlumberger was not only employed to operate the automaton, but also to assist in all aspects of the exhibition, and to act as a kind of clerk to Maelzel—writing out his letters, visiting mechanics on his

An Unreasonable Game

behalf. He was paid $60 a month, the contract for which was drawn up by Charles Schmidt.

Schlumberger was born in Alsace and, after the family manufacturing business collapsed, he earned a living by giving chess lessons at the Café de la Régence, and became one of the best players in France. He was nicknamed "Mulhouse" after the place where he was born, and was particularly good at mathematics. Some said he gambled, others said he drank. Allen does all he can to clear Schlumberger's name. He explains that when Schlumberger and Maelzel had dinner together, they kept a chessboard between them, and Schlumberger was often so immersed in the game that he didn't realize how much he was drinking, and sometimes had to get into the box for the evening performance too soon afterwards. The few games the automaton lost, Allen feels sure, were conducted on these unfortunate occasions.

Allen describes in some detail Schlumberger's appearance: "His countenance was remarkably agreeable in expression; his features well-defined and handsome; his nose well-formed and prominent. The admirable formation of his head, with its dark brown hair, and his beautiful chestnut eyes, are always dwelt upon by those who had known him. His figure was muscular and well-proportioned." Schlumberger died just before Maelzel, in Havana, of yellow fever. The pair had spent so many years together, and had become so close, that many people thought Schlumberger was Maelzel's son.

It is worth noting that not all of the directors were similarly proportioned. Alexandre and Weyle were small, and Mouret, the early pamphleteers would have felt vindicated in knowing, was said to be "practically a dwarf"; but Allgaier was well-built and strong, and Schlumberger, Anderson, and Boncourt were at least six feet tall. Dr. Joshua Cohen, who wrote to George Allen from Baltimore, knew Maelzel and Schlumberger well. (He enclosed with one of his letters a note written to him by Schlumberger, thanking him for his generosity during their stay in Baltimore, and one of Maelzel's business cards.) On Schlumberger's size Cohen remarks: "It was doubtless a great inconvenience to him to be so immured, and we all wondered at the time how he could withstand the intense heat of the exhibition room. It was after such a warm confinement . . . that M. threw open the shutters, and relieved poor S. from his suffering."

The occasion Cohen is referring to is the one on which two Baltimore boys saw Schlumberger coming out of the box: the story had been true, of course, and it was just Maelzel's good fortune that the report was interpreted as a publicity stunt. As Allen put it in his essay, the automaton was frequented by a public who "refused to know a secret which had been exposed and published a dozen times."

The routine inside the box was cramped and ascetic. Little air entered (particularly on hot days or in small exhibition rooms) through the holes at the top of the box, which were concealed by the concave mouldings on the loose-fitting lid. The director worked by the light of a single candle, which stood quite a high risk of setting light to the wooden box at any moment (one possibly apocryphal story has it that Schlumberger was made to dash out of the box and reveal himself because someone in the audience shouted "Fire!"). Two candelabra were lit on top of the box beside the chessboard, no matter what the lighting conditions. One suspicious audience member thought they were there to help a human agent to see, but in fact they shed no light into the box; they were there to mask the smell of burning wax from the invisible light-source inside.

At the start of the show, while the exhibitor opened the small door at the front and then shone a light through the opposite door at the back, the director was scrunched up in the larger cabinet, his knees up against his nose. The exhibitor shut the back door, which was a drop door attached by hinges at the top. He didn't lock it, only—ostentatiously—left his keys in it, and slowly walked round to the front again, giving the director enough time to change his position. As soon as the back door was closed, the man inside the box, who was sitting on a low bench attached to silent rails, would roll backwards into the far left-hand corner (next to the rear door), pushing up with his spine the machinery in the small compartment, which was only false and which folded back into the side. As he reversed, a small opaque partition rose up between the front and back of the small compartment, a cross section of the box—so he only took up the back half, and could not be seen by the audience, even though the front door to that compartment was still open.

This was the crucial part of the exhibition: if the back door had remained open, and the audience could still see through the small compartment, the director would have been seen; but because the audience was at that point directing its attention towards the large compartment, which the exhibitor was about to open, they didn't notice that the impression of the small compartment being completely open was pure illusion. The director quickly turned over a lid that covered his left (so he wouldn't be seen when the automaton's back was shown), pulled back a door to conceal his body to the right (so he wouldn't be seen when the large compartment was opened), and pulled up by a string and fastened with a button a false mechanism (or "trap quadrant," as one director called it) that filled the space in the front part of the large compartment and gave it the illusion of being both uninhabited and mechanically motivated.

This was the director's permanent position for the entire session (it was as Racknitz had drawn it, but with much less freedom to move and in reverse—he would be facing the other way). The exhibitor was now free to open the main compartment, shine a light into the empty space, open the bottom drawer, turn the Turk around to show its back, and begin the game of chess.

During the game the director, tightly packed into the rear half of the box and hunched over a tiny travelling chessboard, had to crane his neck to look at the ceiling of his hiding place, in order to see what move his opponent had made. There were, as Racknitz had explained several years before, metal pointers or discs on the underside of each square; when a piece was moved, the magnet inside it would cause the pointer on that square to swing. On seeing the move, the director repeated it on his own small board so that he could keep track of the game (he played it like this in miniature all the way through), and made his own move using the arm of the automaton. Kempelen's pantograph allowed him to steer from within the box a limb that would make the same movement above it—like puppetry with solid strings. He would raise the Turk's arm, centre it on the desired square, lower it, and, turning the collar at the end of the lever in his hand, make the Turk's fingers grasp the piece. He would then move it to where he wanted it, and release the piece from the Turk's hand.

The pantograph was not the only mechanical device he had to manage. There were, certainly by Maelzel's time, levers to make the figure roll its eyes and nod its head, one that made it rap the table top with its *other* arm (in case of cheating), and a cord that, when pulled, made the Turk say "*échec*." During none of this could the director see outside the box. In the event of a cough or a sneeze, or some such human sound, there was a piece of emergency equipment added by Maelzel: a noisy ratchet set off by a strong spring, which the director could release to camouflage his humanity with a mechanical effect.

There were other emergency measures. The keys were left in the drop door at the back so that if the exhibitor had to communicate with the director, he could go round on the pretext of having to wind up the machine with his keys, lift the door, and speak to him. The door could also be pushed open from the inside. "The director could ask or communicate almost anything," wrote Charles Schmidt to George Allen.

If you have been present at an exhibition you may have seen Maelzel put down a piece repeatedly and quite hard. This arose from the Director's question "What's the move?" because no disk went up. Or you might have seen Maelzel go behind the Turk with his candle, because the Director had communicated the fact that his light had gone out from some accident; he could not take matches on account of being discovered by the smell.

Lloyd P. Smith, later in life a librarian at the Library Company of Philadelphia, told Allen in a letter that as a young man he had operated the automaton after Maelzel's death, at the Chinese Museum. On one occasion, he made good use of the door at the back: seeing that he was about to lose the game, Smith made a loud noise with the fold-out machinery and whispered to the exhibitor, who promptly announced that owing to a mechanical failure the exhibition had to be "unavoidably postponed."

The effects of living inside the automaton must have been many. We know, for instance, that the six-foot Wilhelm Schlumberger developed a pronounced stoop. The chess author George Walker has written of "the Siamese twin-like connection" of the director and the Turk, of "body and board"; and this connection, this encroached intimacy, made the boundary between what was human and what was not indistinct. The Automaton Chess Player embodied a riddle about humanity—could a machine think?—that the existence of a human agent would seem immediately to solve; but the situation may not be so simple. This is in part because of the nature of chess itself: it is not merely a game or a series of calculations, but a mechanical way of thought. As the chess-playing artist Marcel Duchamp would put it a century later, "a game of chess is . . . mechanical, since it moves . . . it is the imagining of the movement . . . that makes the beauty in this case. It's completely in one's grey matter."

There is an object missing from George Allen's collection in Philadelphia. It is either missing, say the librarians, or it is too ephemeral to be found; but it is listed in the original catalogue of the sale in 1878: a ticket for an exhibition of the automaton, signed on the back by Wilhelm Schlumberger. There's something wonderful about this relic—or even the idea of it: it's like an artwork by Duchamp in itself; a found object that is also a clue, a double bluff. The automaton could not write, and could not, except encoded into its style of play, possess a signature. The automaton was famous, but Schlumberger was—at least in this connection—not. Not only was he not famous, he was an impossibility: the automaton was not supposed to have a director—technically Schlumberger was not even meant to exist. But here he was, signing away the secret with a flourish of his pen, as if the automaton could hand out autographs like any other star. And the signature was also a statement: this is who I am.

A story is told about Mouret, the director with whom Maelzel toured Europe. Maelzel, who, as Lewis confirms in one of his letters, was "anything but a good paymaster," had not paid Mouret for some time. They were due to play for the King in Amsterdam, and Maelzel had been given a considerable sum of money for this event. Still he did not pay Mouret. Suddenly, the director contracted a fever, the only cure for which, he insisted, was the payment Maelzel had just received. The exhibitor was distraught—how would they pay for the rest of their tour if he gave Mouret the money? and if he didn't, what would he tell the King? "Tell the King," Mouret shouted, "the automaton has a sore throat."

Both of these instances—the autographed ticket and Mouret's trick—are in their way reformulations of the question that worried the pamphleteers. Can a machine "usurp" human faculties? Can it think? Can it sign its name and thereby guarantee its own authenticity? Can it fail? Can it fall ill? The determined language of the "exposing" pamphlets is rearranged into an object (the ticket), and then into a joke (the sore throat)—they are all different treatments of the same philosophical difficulty, and lead to the conclusion that perhaps the line between man and machine is not drawn straight.

Schmidt, the man who did not, in 1858, want to be identified, began one of his letters to Allen with a mysteriously insistent desire to dissociate himself from the automaton: "As I never really was the director of the Automaton, no more at any rate than Coleman whom you name, or Professor Anderson of Columbia College, both of whom directed it in N. York, I do not see why my name any more than theirs should be mentioned." But Allen did mention both of those other directors in his essay, and held back from naming Schmidt, perhaps because of the strangely personal nature of his introduction. "My connection with Maelzel was a mere boyish frolic," wrote Schmidt. "I was but sixteen at the time, and it originated in my love of chess. I never had any engagement with him, nor would I now care any thing about it, had it not been the cause of very painful and unhappy events in after life to me. I may at some future time tell you how a mere frolic became a source of much unhappiness." Just over a week later, he wrote again, and explained:

> That you may not . . . misunderstand the allusions to myself with regard to my connection with the Automaton, I will state that I was most unjustly charged with travelling with an exhibition, and that this charge led to a quarrel, which never was reconciled, and broke off an intimacy which had become

very dear to me. Were I happy now, had I been successful & prosperous, I would not care what was said
about me, in connection with the Automaton. And
though all this happened many years ago, I shrink
from seeing a statement revived, which at the time
caused me much unhappiness and changed and
destroyed a happy career. Enough of this however.

Schmidt leaves it at that. It is unlikely that, even had he been
able to mention Schmidt, Allen would have made much use of
this vague personal pain in his factual account of the Turk's
travels; but the tone of bitter regret in these letters, their
unspecific wave of explanation, and the final, absolute declara-
tion of a life ruined, offer another kind of information about
the automaton. It was not, evidently, just a machine, or just
an illusion, but at times a kind of parasite. The relationship
between the Chess Player and its director was not merely one-
way, the man giving life to the toy; the toy could also have an
effect on the man's life. When the philosopher Walter Benjamin told the story of the chess automaton as a historical
fable, he wrote that the figure "can easily be a match for anyone if it enlists the services" of the man inside. Put this way,
the tale of the Turk becomes monstrous, not a lifeless thing
directed by a man (or a woman), but the puppeteer *in the service of* the puppet. The anxiety is no longer whether a machine
can be like a man, but whether a man can become a mere
machine.

   If the pamphleteers reflected on the automaton from outside, as uninitiated members of the audience, and the chess
coterie was part of a more intimate group of cognoscenti, then

inside that was an even smaller, more secretive circle—the operators themselves, who were hidden in the automaton. Inside that circle were the minds of the inhabitants, pawns in the mysterious games the machine played with human reason. The Chess Player, which at first seemed to be only one box, was, in terms of the ways in which it was seen and the questions it raised, a series of boxes within boxes, or ever-unfolding secrets. If the automaton set a philosophical problem for those who contemplated it, it could also be said to have set a psychological problem for those who inhabited it.

Jacques-François Mouret, one of the most successful directors of the automaton, was the one who gave away its secret. The great-nephew of Philidor, pupil of Deschapelles, and chess tutor to the family of King Louis-Philippe was, according to his friend De Tournay, "very likeable and gay." He was very clever, sharp, and amusing, wrote the chess author Delannoy, but only his talent for the game "redeemed . . . the rudeness of his manner and a certain licentiousness that he indulged in. He used to be in a continual state of semi-intoxication." William Lewis described him in his memoir as "a jovial amusing man, of slender morals and too fond of drinking." George Walker, in an essay on the chess automaton, tells how drink eventually got the better of him: "He unhappily formed habits of dissipation fatal to his respectability and standing in society." In 1834, poor and desperately needing a drink, Mouret sold the secret of the automaton to the *Magazine pittoresque*. He died three years later, "reduced," says Walker, "to the extremest stage of misery and degradation," having "burnt out his brain with brandy."

Mouret may not have been able to live off his meagre pension from the postal service, or the fees from the chess lessons he gave, but, as his obituary in *Le Palamède* makes clear, he was supported financially by the Parisian chess circle—so the selling of the secret seems less a result of real need than of a compulsion to tell, to be rid of it. The obituary was in part a letter of thanks on Mouret's behalf to those who had paid for his funeral, and explained that the deceased, "one of the strongest chess players in France," had for several years been paralysed in all his limbs. "Death," the obituary read, "came to put an end to his painful position."

This is the legend of Mouret—that, no longer in control of his mind, and having lost the use of his body, he sold the secret of the Chess Player, kept for sixty-five years, for the price of a drink. Of course, Mouret clearly drank before he played inside the automaton—there is no reason to suppose that he became an alcoholic as a result—but his paralysis (as if he were still immobilized in the machine), and the burning out of his brain with brandy (as if setting light to the spirit of reason) have the appearance, in retrospect, of a parable about Kempelen's invention.

Mouret, as the director, was the conductor and victim of an experiment whose object was to ascertain what was the most essential human faculty. What does a machine need in order to be human? And can a human reduce himself to that faculty alone? The more Mouret played, the more uncomfortable he was; his physical self was a thing to be overcome. The life inside the box could only be a life of the mind, since the body had so little room to exist, and because the adverse effects were incremental. Enlisted by the puppet—used—Mouret became, after a while, useless. Another operator took over and Mouret, seeing that only his mind made him human, drove himself out of it. Might he have lost his mind as a result of the knowledge that experiment gave him? Could he, in other words, not have lost his mind but relinquished it—sabotaged it, even?

The supreme intellectual challenge of chess (which made it apparently impossible for a machine to play) is often seen to be a threat to reason. For every chess player there are many lapsed devotees, disillusioned or endangered, unable to deal with its strain. Stefan Zweig wrote his last story on the subject, before committing suicide. Vladimir Nabokov wrote a novel about a chess player who goes insane and, unable to play chess or to live without it, ends his own life. Both these works of fiction offer a picture of what it is like to be wrapped up in chess, to lose one's mind to the game, played mechanically or otherwise.

Zweig's novella, *The Royal Game*, takes place at the outset of the Second World War on a boat from Europe to South America. On the boat is the reigning Russian chess champion, a man without finesse or intellect except in this particular domain. The narrator and a group of other passengers pay him to play against them; just as they are losing, a mysterious man intervenes on their behalf and leads them to a draw with the champion. The man, Dr. B., is urged to play a game one to one, but he says he has not sat at a chessboard for at least twenty years. He explains to the narrator that he came by his instinct for chess another way.

Imprisoned in an empty hotel room by the Nazis for months, with nothing to read or to write with, in a cultural vacuum designed to make him submit to interrogation, Dr. B. had been beside himself. One day he discovered a book in the coat pocket of one of his captors and, hoping it would be Homer or Goethe, something he could memorize and learn from, succeeded in stealing it. To his great disappointment, the book was nothing but 150 championship games of chess, something he had not played since he was a student; but it was all he had. He began by improvising a chessboard and pieces with his bedspread and breadcrumbs he had saved; soon he could see it all in his head. Dr. B. found that learning the games sharpened his mind, but after some weeks he came to an impasse: he knew the games too well, and would have to invent new ones, playing against himself; but the division of his mind into two conflicting camps, "Ego Black" and "Ego White," became too much for him—before long he was trapped in a chess fever, hallucinating, shouting, violent. He woke up in a sanatorium, suffering from what he calls "chess poisoning." On being set free from there and from the Nazis, Dr. B. had been advised never to play chess again. Now, however, on the boat, he consents to one game—for this reason: he wants to "discover whether what went on in my cell was chess or madness."

In Nabokov's novel *The Defence,* Luzhin, a chess champion, suffers a breakdown and is warned that he must survive without chess if he is to live at all. Even when he isn't playing, the world takes on the traits of the game; Luzhin sees chess where it does not exist, he hallucinates it and interprets the world by it. Even as he is jumping out of the window, trying to finally end his madness, he sees the ground divide into black and white squares beneath him: the pattern of the game and of his mind swallows him up.

What Luzhin and Dr. B. suffer from is a syndrome that is

the opposite of being "out of one's mind." They are, in a sense, too far into their minds; what they are grappling with is not madness but a form of impacted sanity. Luzhin's father-in-law calls him a "narrow fanatic," and though the older man presumably means to comment on Luzhin's limited cultural and social sphere, the narrowness of his vision is quite literal. Luzhin and Dr. B. (the latter out of necessity rather than desire) both specialize in playing blind. (In blindfold play a "teller" announces the opponent's move using standard chess notation—the player never feels the board but has to visualize it mentally.) This is chess conducted in the dark, in the mind, like a game between the blind player and his own brain. When these two fictional characters go mad, their blindness flips into a kind of over-vision, in which chess is superimposed on to sight.

In reality, players choose to play blind not simply for show, but because it's like chess in its purest, most concentrated form. This has its attendant risks—for a time, blindfold chess was banned in the Soviet Union, on the grounds that it endangered mental health. The game requires, according to a turn-of-the-century chess manual, "a memory as tenacious as Sir Walter Scott's, the same faculty for picturing what he thought of that Carlyle possessed, the concentration of a mathematician, Caesar's power of quick detachment." Blind chess, in other words, is the game of pure reason.

In 1894, Alfred Binet, the French psychiatrist who invented the IQ test, wrote a case study of blindfold chess players. No one had ever explained in words how they got through this process, and the phenomenon of a game played entirely in the mind remained a mystery. Binet was interested in visual memory and mathematical ability, in whether the players relied on calculation or a sort of "internal mirror." He published a questionnaire in a chess periodical, asking players how they represented the board to themselves—was it in colour? Did the pieces have particular shapes, or distinctions of touch? Was the voice of the teller important, or the personality of the opponent? Several players from the Café de la Régence were brought to Binet's laboratory to play under observation. One player wrote to Binet that blindfold play had "an unfavourable influence on my mental powers." But after all his investigations of the players' brains, Binet concluded that he could find no direct, causal link between chess and madness; there had, after all, only been two famous mad players—Neumann and Morphy. But what had happened to their reason?

Paul Morphy was the subject of a paper given by Ernest Jones before the British Psycho-Analytical Society in 1930. Morphy, who was born the year the automaton ended its tour, had phenomenal skill—he played blindfold games that, according to one admirer, would have had Philidor turning in his grave—but after a triumphant tour of Europe at the age of twenty, during which he beat the leading player in the world, Morphy returned to his home in New Orleans and retired from professional chess. He attempted to take up law, but had little success, and eventually retreated into seclusion and paranoia. He believed that his brother-in-law was trying to steal his inheritance, poison his food, and destroy his clothes. On one occasion he attacked one of his best friends, whom he suspected of wanting to kill him. He would pace up and down on his veranda like a latter-day Don Quixote hallucinating chess moves, declaiming in French: "He will place the banner of Castile on the walls of Madrid to the cries of the conquered city, and the little king will run off in shame!" The mere mention of the game provoked in him "strong revulsion." He died at the age of forty-seven, of "congestion of the brain."

Jones's purpose is to investigate the relation of this "mental catastrophe" to the game of chess. "It is impossible to believe," he writes, "that there was not some intimate connection between the neurosis . . . and the superb efforts of sublimation which have made Morphy's name immortal." A serious game of chess, he suggests, places a good deal of strain on "psychic integrity." Jones traces the etymology of chess, from the Sanskrit *chaturanga* to the Persian *chatrang* to the later Persian *Schah*. So checkmate, or *Schah-mat,* means, literally, "The king is dead," or, in a slightly different version, "The king is paralysed, helpless, and defeated." Not only is chess a substitute for war, Jones argues, but it is also motivated by the desire to murder the father, or kill the king. He places great emphasis on the sudden death of Morphy's father, the year before he won a big American chess tournament. As the game evolved through the centuries, the piece known as the queen was given more power, thus, Jones explains, completing the Oedipal formula ("It will not surprise the psycho-analyst when he learns the effect of the change: it is that in attacking the father the most potent assistance is afforded by the mother.").

Jones mentions a famous match that Morphy won on the basis of a series of imagined future moves so complicated and numerous that, as an analyst of the game later claimed, no human brain could calculate them. Elsewhere, he cites an occasion on which Morphy played for ten hours straight without taking food or water, and repeated instances of his having played from nine in the morning until midnight without the slightest signs of fatigue: evidence of supreme sublimation. Sublimation, Jones explains, has a defensive function—so chess, you might say, protects the player from the real world. To put it crudely, when these sublimating powers break down—when, for whatever reason, the world of chess is smashed—the player is left to face whatever it is he was avoiding through the game.

One of the most puzzling questions about the automaton is why such virtuoso players should have chosen to operate it; and the answer may be this: that, like playing blindfold, playing inside the box accentuated chess—it was like chess distilled, the discomforts of the body becoming sublimated into the game. If the object of chess was to "paralyse the king," then Mouret, stuck inside the dark chest of the automaton until seized by paralysis, could be said to have been conducting a parallel game within himself—he became a living re-enactment of the game that was his life, and he died as its fatality.

William Lewis's manuscript contains a wonderful story about an occasion when he was inside the automaton. It shows that although the automaton's game was not literally blindfold (Lewis could see a replica of the board, and did not have to memorize the moves), there were other parts of the operator's vision that were blocked; but his perception of the personalities expressed through chess moves was so acute that he could guess, from within the box, the identity of his unseen opponent. Chess was not about seeing at all. This is Lewis's description:

One evening a gentleman came fully determined to win & accompanied by several friends who believed he would; I opened the game with the King's Gambit, my usual beginning as leading to brilliant moves. I found my adversary played all the correct moves even against one or two which were not in the books; I began to consider who this could be & determined in my own mind that it must be Peter Unger Williams, a first rate player who for some time had been absent from London and was therefore out of practise; if this were so, it behoved me to play very carefully; after some moves I won back the pawn with a good position and after playing an hour and a half I had so improved my game that I had a passed pawn and a winning position and he therefore resigned. My conjectures about my adversary were correct, it was Mr. Williams, an old pupil of Sarratt's who learned the moves and became a first rate player in 18 months, often boasting that he was the best player in England . . .

When I told M[aelzel] about Mr. Williams, he with his usual tact & judgement saw that it would be wise to enlist him as a friend, and having called on him it was arranged that instead of opposing the Aut. he should occasionally assist me in directing it. This suited me very well & I must say for Mr. W that the Aut. lost nothing of its reputation when he directed the moves.

So playing inside the automaton was, like blindfold chess, a supreme challenge, a test of the mind. To audiences the

An Unreasonable Game

Chess Player was a machine that could think, and for the operators inside, it required a superior effort from their own brain. As Simon Schaffer has put it: "For some it was a machine that displayed remarkably human attributes, while for others, including its designer, it was a human who performed in a strikingly machinelike manner." Was this, in fact, the end result? Did Mouret become a machine? Was he made mechanical, by giving over his intelligence to chess?

Over a century and a half after Mouret's death, the world chess champion Gary Kasparov lost a game to a computer, Deep Blue. Years later, he still maintained that the machine had been tampered with. I met him briefly in London, over a game of chess played on computer screens (which I quickly lost), and I asked him, uncomprehending, what could possibly have been done to Deep Blue to make the machine beat him. Kasparov became extremely agitated. "No!" he shouted, waving his fist. "No! Machine cannot beat me!" It was the combination, he explained, of man and machine that had won: somewhere along the line, he believed, a human being had intervened in the computer's calculations, and these joined forces, this man-machine, was all-powerful—or certainly too much for him.

Whether or not Kasparov is right about the subterfuge, his perception of the ideal combination is intriguing. In the world of men and machines, men seem to become shrivelled, like the automaton's operator as described by Walter Benjamin, "wizened and . . . out of sight." In the years after the automaton was destroyed, men and women were to feel increasingly like machines every day—even more than in the era of Vaucanson's silk workers, the late-nineteenth-century world of work

109

would transform people into cogs and wheels, inhuman-seeming parts of mechanical factories. Yet before their bodies melded with machines, Kempelen's Chess Player provided a riddle about their brains. Can a machine think? was the question. Only, you might say, if a man relinquishes the ability to think. If chess—particularly blindfold chess, chess in the mind—can drive its players mad, then the automaton was an accentuation of that. In the game of artificial intelligence, the only true loser may well be human reason.

12/13/2007

CHAPTER THREE

Journey to the Perfect Woman

Those mechanical wonders which in one century enriched only the conjuror who used them, contributed in another to augment the wealth of the nation. Those automatic toys which once amused the vulgar, are now employed in extending the power and promoting the civilization of our species.

—David Brewster, *Letters on Natural Magic*

She was created to be the toy of man . . . and must jingle in his ears whenever, dismissing reason, he chooses to be amused.

—Mary Wollstonecraft, *A Vindication of the Rights of Woman*

It used to be called the Invention Factory, but now it is a museum, a frozen piece of industrial life, all wheels and pulleys and vices and clocks. Thomas Edison's laboratory in West Orange, New Jersey, was built in 1887, and planned, as Edison put it, as "the best equipped and largest laboratory extant." It took up 22 acres, and was designed so that the ideas of "the Inventor of the Age" could be put into mass production as soon as he had them. Inventing would no longer be the province of the lonely genius, Edison boasted; in his hands it was to become an industry. Gone were the days of the marvellous curiosity, unique creations such as Vaucanson's duck and Kempelen's chess automaton; now wonders would be churned out by the thousand, punched and pummelled by heavy machines, shunted down the production line.

Half of Edison's 1,093 patents were obtained from here, and he told his staff of 10,000 that he expected a major invention every six months, and a minor one every ten days. Despite the factory ethic, these wizardly activities were conducted in the utmost secrecy: the grounds were surrounded by a high picket fence, and were so secure a novice gatekeeper refused to allow entry to Edison himself until he could get someone to identify him.

By the time Edison commissioned this set of buildings, and moved here from his lab at nearby Menlo Park, he had become the most famous man in America, a person on whom were pinned all hopes for the future. He had invented the electric light bulb, he had succeeded in capturing human speech on a cylinder of tinfoil, he had improved the telephone—who could tell what he would come up with next? He became a kind of mythical figure, much reported in the press. The *New York Daily Graphic*, which had given him his best-known nickname, "the Wizard of Menlo Park," hired a reporter to visit him once a month, just in case a new invention had passed them by. Edison liked giving interviews; as his biographer Paul Israel points out, he "actively worked to advance his image as a modern-day Prometheus who had single-handedly transformed the world with his inventions."

Apocryphal stories seemed no more unlikely to the public than the real ones. Edison's private secretary, Alfred Tate, had form letters which he sent out in response to the hundreds of zany queries: no, Edison had not created a new star by sending an electric light into the sky attached to a balloon; no, he had not patented the 365-day shirt, which had an outer layer for each day of the year, but an unchanging inner layer. Still, anything, when associated with Edison, seemed possible. Early on in his career, one of his backers wrote to him saying, "If you should tell me you could *make babies by machinery*, I shouldn't ⌐ doubt it."

Much of the West Orange factory has been left as it was: the elegant, heavy, red-brick main building, 250 feet long and three storeys high, and a smaller single-storey building, left open as a sample of others now burnt down or closed—the building used for kinetograph experiments, for example, or the phonograph works, destroyed by fire in 1914. The smaller building houses Edison's chemistry lab, full of row upon row of powders and pumps. His white lab coat has been left there, and on the wall is a tall, thin branch, a sample of "goldenrod"—a type of plant researched and cultivated by Edison, primarily in the botanic gardens of his winter home in Florida. In the 1920s, fearing a war might cut off foreign supplies, Edison's friends Henry Ford and Harvey Firestone asked him to investigate possible domestic sources for a product that had become indispensable to their automobile industry. So Edison went on to make an unwitting contribution to the afterlife of Vaucanson's blood machine: goldenrod was discovered to be a perennial, domestic, abundant source of rubber.

At the entrance to the main factory building, the preserved time clock, where Edison used to punch in and out like the rest, hangs on the wall, with useless cards still slotted into place. The corridor ahead leads to the stockroom and machine shop, and a double door on the right opens on to Edison's grand library. It's an extraordinary room—three floors high, with two galleries of bookshelves. Edison's desk is here, and a camp bed remains unmade in an alcove; this is where he would nap if he was working several days and nights in a row. (Edison's ability to fall straight to sleep anywhere, even standing up, was well known. Alfred Tate was so taken with this knack he thought Edison might have "invented sleep.")

Amidst all the usual accessories of a library or office—a conference table, leather armchairs, a hatstand, engravings—are souvenirs of Edison's travels in thought and space: the Edison Vitascope, an early projector for moving pictures, and a marble statue called "The Genius of Light," which he brought back with him from the Paris Exposition of 1889 (the centenary celebration of the French Revolution), where he had dined with Gustave Eiffel in his brand-new tower. The sculpture was made by an Italian, and shows a winged boy sitting on the ruins of a gas street lamp and holding a bright electric light bulb up in the air, like a budding Prometheus with a new form of fire. The allegory represents the triumph of electricity over gas, and proves, since it wasn't commissioned, how widely glorified the products of Edison's mind were.

The library was, in the late nineteenth century, one of the best scientific and technical libraries in the world, and covered botany, astronomy, medicine, and philosophy, as well as the more expected subjects of mining or mechanics. It held literature in a number of languages (Edison had a translator on his full-time staff so that research could be conducted here by anyone in his employ), and if any book or journal was needed in the course of a particular project, Edison would order it immediately. Not only books were kept here: two sides of each gallery were made up of glass-fronted cabinets, like old-fashioned cabinets of wonders, which contained specimens of rare minerals from all over the world, tagged and numbered, and miscellaneous anatomical models from Edison's various autodidactic enterprises.

Edison's official biographers, Frank Dyer and T. C. Martin, claimed while Edison was still alive that he was "the living embodiment of the song *I Want What I Want When I Want It*." However this poetically titled song went, Edison certainly met his needs in ordering up the contents of his stockroom. Protected even in Edison's time by a long, tall, metal grille, with a large arched window lighting its many drawers and shelves, the dust-filled stockroom is like a cage for magical materials. The museum guide claims that anything could be found here, "from the hide of an elephant to the eyeballs of a U.S. senator," and certainly Edison demanded that it contain, as he wrote, "a quantity of every known substance on the face of the globe . . . all kinds of ores, metals, fabrics, gums, resins, and samples of every imaginable material." This included animal horns, teeth, skins, shells and feathers, grains and roots and woods, papers of all weights and colours, and chemicals, like the fluorescent salts Edison had made himself. There were also the tools required to mould these materials, in large quantities and still arranged on the open shelves—screws, bolts, hammers, drills, needles, twine, and all kinds of other pieces of machinery. Edison wanted to be able to build "anything from a lady's watch to a locomotive" here, and he couldn't be running out of material at the moment of inspiration.

The neighbouring machine shop is the most magnificent

relic—not because it is grand, as the library is, but because this room, with its rows of weighty steel wheels and rolling pulley-belts overhead, is like the engine room of heavy industry, a picture of how that masculine world was set in motion. Everything is hard and black and big, laid out for repetitive, mechanical work: it is the world that took over from Vaucanson's system, and paved the way for the Model T. Suddenly, machines and motion were paramount; there are no people here, and that seems right, since even when there were they were only meant to be machines, or parts of machines. In a corner, on a rack by the long, broad sinks, factory workers' coats and hats are still hanging up, like the strange ghosts that technology left behind.

In the second half of the nineteenth century, work had been transformed. The informality and craftsmanship of the workshop, along whose lines Edison had still operated at Menlo Park, gave way to the precision of the factory (although where Edison was concerned it was not a straightforward shift: he embraced industrialization whilst holding on to the myth that he alone was responsible for his factory's output). Time clocks like the one at West Orange, the first physical sign that workers were to be defined only in terms of numbers and hours, were an invention of the 1890s. In the experimenting arm of his laboratory Edison stuck to his old ways—he had a team of loyal experimenters (some of whom left when they received no credit for their inventions), but in terms of production, his factory operated on what was called "the American system," a precursor of Fordian mass production in which each worker performed a single, simple, repeti-

ie, the invention of a machine was, like a machine, the thing he opened.

tive task all day, which required no specialization and allowed no variety. The principle of this system was that if work previously done by one person was subdivided into several parts, output would be greater. So, the work of a single craftsman could be divided into seventeen or more different occupations. No job took long to learn, and specialists were out of business. A new working class was created in America, of men and women, paid next to nothing, who were cogs in the new industry. By 1900 they made up a third of the country's population, and lived below, or just above, the poverty line.

The point about this is not just that conditions were bad—although they were, and there were an unprecedented 37,000 strikes in twenty-four years, including one at Edison's Machine Works, when an end to piecework was demanded and never granted—but that the idea of humans being turned into machines was becoming a current topic for political discussion. It was what Karl Marx had written about in the *Communist Manifesto:*

> Owing to the extensive use of machinery and to division of labour, the work of the proletarians has lost all individual character, and consequently, all charm for the workman. He becomes an appendage of the machine, and it is only the most simple, most monotonous, and most easily acquired knack, that is required of him . . . Masses of labourers, crowded into the factory, are organized like soldiers . . . As privates of the industrial army . . . they are daily and hourly enslaved by the machine.

Considerable attention was given to this problem in America. Bishop Henry C. Potter said in 1897 that the strikes and the saloons were understandable responses to what he termed the "mechanicalization" of the workman. Another clergyman wrote that the factory worker was nothing but "a tender upon a steel automaton," and that the piece system offered no satisfaction: "He sees no complete product of his skill growing into finished shape in his hands. What zest can there be in the toil of this bit of manhood?" And in England John Ruskin complained that mechanization had turned men into "mere segments of men." In other words, the problem was not just that workers only made a small piece of the final object, but that in the process they themselves were broken into pieces, as if producer and product were so closely identified with one another that they took on each other's attributes, and as if man, in making machines and operating machines, must inescapably lose his "manhood" and become a part of a machine himself.

It is a scenario familiar from early-twentieth-century films such as Fritz Lang's *Metropolis* and Charlie Chaplin's *Modern Times*, but in late-nineteenth-century America nowhere was the mechanization of human beings more ironically in force than in the production, piece by piece and thousand by thousand, of Edison's talking doll.

Only a month after he had first invented the phonograph, in 1877, Edison had already made arrangements to manufacture a smaller version for use in clocks and toys. At the Paris Exposition of 1878, where he astonished the world with his talking machine, and was awarded the Grand Prize for being "the Inventor of the Age in which we live," he exhibited this miniphonograph, and a design for a doll from which sound would emerge via a funnel sprouting freakishly from the top of its head. Since this doll looked, as *Harper's Young People* reported, "more like a miniature stove-pipe," and since "Mr. Edison knew more about phonographs than children's nurseries," this particular structure was abandoned; but ten years later, when he returned to the phonograph after inventing the light bulb, Edison also revisited the idea of a talking doll. The contrast between the doll and the heavy industry of the Edison factory seemed very appealing. Edison's colleague W. K. L. Dickson later wrote that it was "perhaps the daintiest and most suggestive of all the multiform uses to which the phonograph has been put." He described "roseate lips," which would "lisp out the oft-conned syllables of nursery rhymes, pipe the familiar strains of Mother Goose's ballads, and give forth the cooing and wailing sounds of baby life . . . Under such auspices," Dickson asked, "into what enchanted realm will our ordinary toys be transformed?"

The truth, of course, was not really dainty or enchanted. The fact was that dolls were big business in Europe, but the Americans had lagged behind, making them traditionally in their homes or importing them, at great expense, from France or Germany. The new inventors of the nineteenth century wanted to put this right, and they brought all their industrial skills to bear on the matter. Of course Edison knew more about stove-pipes than nurseries (although he did, by then, have three children of his own, and would have another three later, when he was putting his doll into production), but so did all the others who contributed to the doll-making business in America. Those who were newly employed in doll manufacture had previously specialized in printing or typesetting, in making clocks or pulley-belts. Cab Ellis learnt how to make mechanical joints for dolls when he patented his "steam excavating machine" for railways. Vincent Lake had just finished his typographic machine when he patented his "All Steel Doll"—and Edison was no different. When Albert Hopkins, the editor of *Scientific American*, came to visit Edison's factory and wrote a cover story about the doll in April 1890, he noted admiringly that its manufacture "calls into requisition the skills of mechanics in almost every branch, and it has necessitated the construction of new tools which are interesting of themselves."

Edison's doll, now barely a footnote in biographies of the inventor, was in 1890 no small affair. In fact, Edison devoted so much time, space, and manpower to this doomed project that one might interpret its disappearance from the story of his life as the cover-up of some shameful secret—was it a project akin to Vaucanson's abandoned blood machine? It is as if Edison's large-scale attempt at creation was bound to fail, and had to be buried.

Hopkins recorded in his article the existence of "an extensive building exclusively for the manufacture of talking dolls," and this is confirmed by a notice in the local newspaper, the *Orange Chronicle* of 1 February 1890, which reported a new building for the doll-makers measuring 40 feet by 210 feet. Another building, it claimed, was also scheduled to be built. Five hundred people were employed at the phonograph works, and it took half of them to make each doll. There was a rigorous production line, and, Hopkins wrote, "order and system reign[ed] in every department."

The dolls' bodies, which were made up of six different pieces of tin, were stamped out and soldered together by steel dies and iron presses, some of which weighed five tons. The miniature phonographs were made in a separate assembly room, and adjustments to them were made in another. In one room eighteen young girls, each with her own cubicle, sat speaking into the machines, recording the words the dolls were to say. Hopkins was quite overwhelmed: "The jangle produced by a number of girls simultaneously repeating 'Mary had a little lamb,' 'Jack and Jill,' 'Little Bo-Peep' and other interesting stories, is beyond description," he wrote. "These sounds united with the sounds of the phonographs themselves when reproducing the stories make a veritable pandemonium." When the girls had finished, the phonographs were taken into another room, where they were put into the bodies of the dolls, and from there the dolls passed into the packing room, where they were put into boxes labelled with "the story the doll is able to repeat." In other words, it was a serious operation; Edison's factory had the capacity to make 500 dolls a day—that is, over 100,000 dolls a year.

Far from the dainty image offered by Dickson, these dolls did not inhabit an "enchanted realm" of "roseate lips." They were produced at speed, on an industrial scale. The expenses in Edison's accounts for "Sundries in connection with Experiments on Toy Phonograph" list components such as might be used to make some heavy piece of equipment, rather than a child's toy. "70 lbs. of cast iron," the book reads, amongst other entries for March 1889, "14 oz. of 3/4 inch brass rod, 6 11/4 18/18 R.H.I.M. screws, 3 1/2 oz. #7.8 & 10 microscopic glass, 3 3/4 oz. Stub steel." In December the experimenters added sheets of tin and zinc; at other times they ordered black rubber tubing, asbestos board, and steel piano wire. Every now and then a mysterious demand would crop up, reminiscent of the gilded female robot in *Metropolis*—an ounce of celluloid, for example, half a pound of dental plaster, or a couple of ounces of gold lacquer—but for the most part, the materials ordered up offer an image of this small child being moulded from hard-core ingredients.

Hopkins's reservations about the figures' speech were echoed in *Harper's Young People*'s description of their "tiny Punch and Judy tone," and were later reflected in the views of the public, who complained on one occasion that "the voices of the heavy little monsters are exceedingly unpleasant to hear." The horror seemed to stem from the close physical resemblance of these products to human beings. Little talking girls were spewing forth from Edison's factory, as if they were lamps or clocks. The historian Miriam Formanek-Brunell has quoted a visitor to another doll factory, and his reaction might just as well be applied to the scene at West Orange. "There is a special machine for stamping out the hands," wrote this reporter from *Toys and Novelties*:

> I stood in front of it, fascinated by the steady stream of queer, little hands that fell ceaselessly from the iron monster—it was awful, uncanny, hypnotizing. Indeed, the whole sight was grim and monstrous. The low factory rooms were misty with steam and lit by strange, red-glowing fires; always the steel machines pulsed and clanged; and through the mist sweaty giants of men went to and fro with heaps of little greenish arms and legs.

The dolls themselves were not the most child-like of things. The metal body, a mannish piece of muscular armour, had been perfected over the course of thirteen separate patents, the dates of which were engraved on its interior. It had a perforated chest, so that the sound could escape. There was a door in the back, which could be removed if the phonograph inside was out of order, or if the owner wanted to replace it with another phrase. There was a handle sticking out of its spine, which had to be cranked in order to make the doll talk. It was attached to a spring, which would return the phonograph to its starting place as soon as it had finished, so that the phrase could be repeated instantly, again and again. The head of the doll, with its painted eyebrows, glass eyes, and fierce little teeth, was made in Germany, and the jointed wooden limbs and composition hands and feet were made elsewhere in the United States. All of this was brought together at the West Orange site, to make a 22-inch, 4-pound doll. It cost $10, more than a week's wages for most, and it was too heavy for young children to pick up with ease. Despite his children, Edison did indeed know little about the nursery. One can only conclude that the dolls were not for children, and adults like Hopkins and the *Toys and Novelties* reporter were not alone in picking up on their aggressive horror. Formanek-Brunell quotes a survey taken at the time Edison's dolls were manufactured, in which a four-year-old girl, fusing the animate with the inanimate in a way that recalls Vaucanson's duck, said she didn't like talking dolls, because "the fixings in the stomach are not good for digestion."

Perhaps one of the first and most remarkable things to

notice about the talking doll is the speed with which Edison conceived it once he had invented the phonograph. Edison called the phonograph "my baby," and W. K. L. Dickson claimed that "no child of Edison's brain has ever received such fostering care." The phonograph was already imagined, that is, in human terms. "Mary had a little lamb," famously, were the first words spoken by the phonograph, in Edison's own voice. It began by speaking the words of a child, and it was not long before a child was invented to give it shape, or to give it life. So the capturing and reproduction of speech were accompanied by a casing for it in human form.

It was not the first time a speaking machine, or a speaking child-figure, had been attempted. The very first talking doll was patented, in fact, by Johann Nepomuk Maelzel in 1824, while he was on tour with the Chess Player. He designed a pair of bellows that, when attached to a tube, a widening oral cavity, and a set of valves, could say "*papa*" or "*maman.*" This was the system Maelzel adapted in order to make the chess automaton say "*échec.*" In designing his doll, Maelzel brought to fruition a project that had long preoccupied his predecessor.

After he had constructed his Automaton Chess Player, Wolfgang von Kempelen turned his attentions to a machine his friend Gottfried von Windisch described as "much more worthy of admiration." Kempelen made a series of metal jaw-like structures, each of which could imitate the sound of a vowel, by receiving air through a tube, according to a principle similar to that of an organ. But his initial experiments were far from perfect; he found that the closer he came to simulating human features with his machine, the better the sound was. So he restricted himself to one mouth, and one glottis, made of elastic gum, and he added a nose made of two tin tubes, through which he could achieve the effect of consonants.

Kempelen was proud of his invention. In a letter to Benjamin Franklin of 1783, he invited the distinguished American to visit the Chess Player and this "other very interesting machine which I am working on" in Versailles. It had, by all accounts, a fairly recherché repertoire of phrases. David Brewster cites the following as its specialties: "*opera, astronomy, Constantinopolis, Vous êtes mon ami, Je vous aime de tout mon Coeur, Venez avec moi à Paris, Leopoldus secundus, Romanorum imperator semper Augustus,* &c." Thomas Collinson reported that on two occasions (in London and Vienna) he had asked the machine to say the same word, "exploitation," which it did, "distinctly pronounced with the French accent." Windisch even remarked on her personality traits. He wrote that the voice of the machine (the word is in the feminine in French) "is pleasant and sweet; she pronounces only her 'R's with a rasping noise and a slight snort. If her answer has not been understood, she repeats it, more slowly; and if she is asked again, she repeats it once more, but with a moody and impatient tone."

Kempelen never finished this creation. It remained in the shape of a simple box, although, according to Windisch, he had other plans. In 1783, he wrote:

> This speaker does not yet have a human body . . . In order to lessen the volume of his luggage for the trip, the Author has postponed until his arrival in Paris the external dressing of this Machine. He plans to give it the appearance of a five to six year-old child, because she has a voice comparable to that age, which is in any

case more in proportion to the actual age of this Machine, still far from the point of perfection. If she happens to mispronounce a few words, she will, with the looks of a child, more easily come by the indulgence she requires.

It was an extraordinary scheme, an intentional confusion of man and machine. Indeed, Kempelen's illustrations of this machine continually meld the human and the constructed. They are like drawings from textbooks on both medicine and mechanics, superimposed on to each other. The way air is exhaled is demonstrated by a pair of bellows in place of lungs, operated like an accordion by human hands; they are joined at the top by a tube, which rises like a trachea and has as a kind of lid a human tongue. An inset drawing reveals two tubes, which could be arteries, but might equally be steel pipes. In another plate, the bone structure of a human jaw, with teeth graphically portrayed, has bursting out of it a gust of air, or speech, or blood, like a man-made geyser implanted into a cheek.

In conceiving the uncanny costume for his machine, Kempelen imagined a child to embody all the perfection of an unsullied being, who is therefore forgiven anything; but this child simultaneously plays the part of imperfection, of the not yet fully formed and fallible human. In practice, however (and this is the crowning part of the multiple bluff), the child is not a person at all but a machine, which, man-made and theoretically infallible, will camouflage its imperfections by taking on the mask of a child, thereby making itself seem all the more perfect—that is, all the more life-like a simulacrum of a living thing.

This plan, though never executed, makes Kempelen and his intended child the direct precursors to Edison and his phonographic doll; but the traditional history of the phonograph would have it another way. In the year the phonograph was first exhibited to the public, *Popular Science Monthly* explained that speaking contraptions had until then aimed to reproduce the mechanical causes of the voice—that is, they had tried to emulate the human organs that produced speech. Edison, on the other hand, made no attempt to construct an android and, ignoring the causes of speech, reproduced its effects instead. The phonograph, in short, was remarkable partly because it did not look human—it spoke just like a person, but looked like a machine, a simple cylinder of tinfoil. It possessed none of the properties by which Vaucanson had made his name with the Flute Player, constructing a bellows or lever for each muscle or vein, concealing a metal tongue and leather fingers. As Edison himself put it, "It is funny, after all. You have to pucker up your mouth to whistle, but the phonograph doesn't pucker one bit."

Popular Science Monthly contrasted Edison's invention with a famous android, or semi-android, that had done the rounds in exhibition rooms in Europe some decades earlier. "Euphonia," as she was called, was the work of Professor Faber, a German astronomer. One observer described Faber as a "sad-faced man," travelling with "his child of infinite labour and unmeasureable sorrow." John Hollingshead, a theatre manager and occasional contributor to Charles Dickens's journal *Household Words,* recalled the melancholy scene in his memoirs. "Its mouth," he wrote, "was large, and opened like the jaws of Gorgibuster in the pantomime, disclosing artificial gums, teeth, and all the organs of speech." Hollingshead called Euphonia the exhibitor's "wonderful toy," his "scientific *Frankenstein* monster," and remarked with a strange insistence that he was certain the two slept in the same room. A "hoarse sepulchral voice came from the mouth of the figure," he wrote, "as if from the depths of a tomb." The exhibition was not much appreciated, he claimed, and when Faber tragically "destroyed himself and his figure," the world went on just the same. "As a reward for the brutality," Hollingshead concluded, "the world, thirty years afterwards, was presented with the phonograph."

This was the machine that was said by the scientific press to be the phonograph's opposite; but was it really, given Hollingshead's conclusion and Edison's desire to insert the phonograph into a doll, that far away? In fact Edison did disclose to one newspaper that he was thinking of giving his machine more resonance by adding "a voice chamber over the mouthpiece about the size of a human mouth, with teeth and perhaps a tongue." It was as if he couldn't resist granting his creation a human form; as if, perhaps, the pure simplicity of the mechanism was too strange. Although he always said the phonograph was his favourite and the most perfect of all his inventions, Edison did admit to being "a little scared" when it worked the first time; and the "sepulchral" impression of Euphonia that Hollingshead reports was shared by the earliest audiences of the phonograph.

Death seemed to be on everyone's mind, observing Edison's invention—either in terms of the machine itself ("we must raise our estimate," declared *Popular Science Monthly*, "of the powers and potencies of 'mere dead matter' "), or the metaphors used (Edison's colleague W. K. L. Dickson referred to the product as "sounds embalmed"), or even as a thing to be overcome. In many cases, the possible "return of the dead" was broached early on: "What a pity you hadn't invented it before," a *New York Post* reporter told Edison. "There is many a mother mourning her dead boy or girl who would give the world could she hear their living voices again—a miracle your phonograph makes possible." Most considered this future outcome in amazement: "Of all the startling powers of the phonograph," ran an article in as technical a journal as *Engineering*, "there is none perhaps so extraordinary as its capability of reproducing, years after, the voices of those who are no longer on the earth." But others were unsure, more anxious, perhaps, about where this would lead, as the open-ended phrasing of *The Times* suggests: "What will be thought of a . . . mere machine by means of which the old familiar voice of one who is no longer with us on earth can be heard speaking to us in the very tones and measure to which our ears were once accustomed?"

Most dramatic of all was the exalted speech of Edison himself, as he explained to the reporter from the *New York Post*,

> Your words are preserved in the tin foil, and will come back upon the application of the instrument years after you are dead in exactly the same tone of voice you spoke them in . . . This tongueless, toothless instrument, without larynx or pharynx, dumb, voiceless matter, nevertheless mimics your tones, speaks with your voice, utters your words, and centuries after you have crumbled into dust will repeat again and again, to a generation that could never know you,

every idle thought, every fond fancy, every vain word that you choose to whisper against this thin iron diaphragm.

It was as if, with these declamatory words, Edison was setting down, to the awe of his contemporaries and for the record of the history books, his ability to play God. He was the creator, as he states quite clearly, of something that would outlive human beings, something that could capture them, as it were, that could "embalm" their voices and remain immortal. Indeed, *Engineering* magazine commented in 1878 that, when faced with the phonograph, "it is impossible altogether to resist a feeling of wonderment, recalling to one's mind perhaps the feelings of Pygmalion or the hero of *Frankenstein*."

Edison was thought to hold the key to technology to such an extent that he was asked on a number of occasions to speculate about the future. He agreed to collaborate on one project, a science-fiction novel, to be co-written with George Lathrop, Nathaniel Hawthorne's son-in-law, and provisionally entitled "Progress." He was to provide the ideas; Lathrop would turn them into chapters. But after handing over one set of notes, Edison went back to more practical work. As he brought out a new invention, he would speak about it to the press, and Lathrop soon felt betrayed. How could Edison tell everyone about these machines, he wondered, when he should be saving them for science fiction? "The Kinetograph has already been 'given away,'" he wrote to the inventor, "and soon there will be no novelty left to describe in my story."

What is wonderful about the complaint, and the subsequent abandonment of the project, is that Lathrop implies it

1. Vaucanson's automata, as exhibited in 1739: the pipe and the drum player, the digesting duck and the flute player. This engraving, by the distinguished eighteenth-century book illustrator H. Gravelot, accompanied the original pamphlet in which Vaucanson explained the inner workings of his inventions.

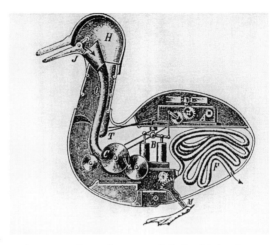

2. The purported guts of Vaucanson's duck. This drawing was made by an American inventor, in an attempt to guess how the duck managed to digest its food. However, later mechanicians who repaired the duck suggested it didn't work this way at all, and that there was some cheating involved in the machine.

3 and 4. These two photographs, taken in the 1890s, are thought to represent the skeleton of Vaucanson's duck, and the large pedestal in which its mechanism was hidden. Nothing more is known about the machine.

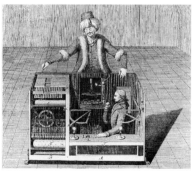

5, 6 and 7. The front and back views of Kempelen's chess player (*top and middle*) were published in a pamphlet written by the inventor's friend Karl Gottlieb von Windisch. They were designed to show that no man could possibly be hidden inside the machine. The third view was an attempt to expose Kempelen, drawn by Joseph Friedrich Freiherr du Racknitz, who guessed, incorrectly, that a dwarf was hidden inside the box.

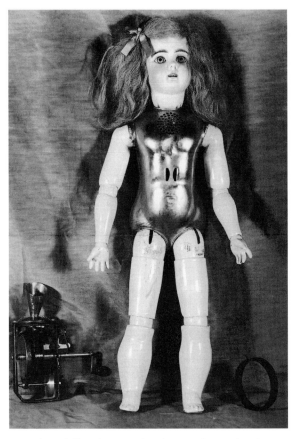

8. A talking doll made in 1890 by Thomas Edison. Edison's wizardly projects provided the inspiration for a novel, *The Eve of the Future*, in which the inventor constructs the perfect woman. The doll was manufactured by the thousands.

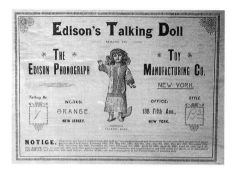

9. (*above*) A label from the packaging of a talking doll. The "talking number" indicates which rhyme she could recite.

10. (*right*) An engraving of Edison's talking doll, as printed on the cover of *Scientific American* in April 1890. An adult hand, just visible, turns the crank in her back to wind up the phonographic mechanism.

11. The manufacture of talking dolls at the Edison factory in New Jersey. The phonographs are inserted into the dolls' bodies, then the dolls are dressed and packaged in boxes. The walls are lined with shelves and filled with dolls.

12. (*left*) The magician John Nevil Maskelyne, pictured with his automaton, Psycho, on the cover of a pamphlet advertising his performances at the Egyptian Hall in 1875.

13. (*above*) A still from Georges Méliès's 1902 trick film *The Man with the Rubber Head*, in which Méliès pumps up his own head until it explodes.

14. Georges Méliès, living in oblivion towards the end of his life, and working in a small toy stall in Montparnasse station.

15. The Doll Family in the 1930s with the silent movie star Harold Lloyd. *From left to right:* Tiny, Harry, Daisy and Grace.

16. The Doll Family in the 1920s, dressed in the glamorous outfits they wore for their performances with the Ringling Brothers Circus.

17. Harry Doll with Jack "Sky High" Earle, a fellow Ringling Brothers performer. The two of them liked to greet each other loudly as "Daddy" and "Junior" around the towns they toured.

18 and 19. Front and back views of Jacquet-Droz's eighteenth-century writing automaton in Neuchâtel.

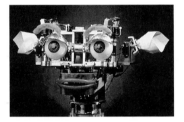

20. Kismet, as seen at the humanoid robotics lab at the Massachusetts Institute of Technology.

21. Hadaly, at the Takanishi lab at Waseda University in Tokyo.

was difficult to distinguish, in this age of invention, between sci-fi flights of fancy such as the ability to separate one's own atoms at will, and current technological advances like the movie camera. All of it, possible or impossible, was in Edison's hands, or in his brain. In a sense he was a truly Promethean figure: at a time when machines were taking over the labour force and doing away with the need for people, Edison was the master of mechanism. He could override mortality, he could animate "mere dead matter"; he had become, in the eyes of the world and perhaps even in his own, the author of the future.

So whether or not the phonograph initially had the physiological properties of a human being, Edison evidently felt in charge of life itself. He was ready to construct a perfect man, or, as it turned out, the perfect woman.

In 1878, the year Edison exhibited his newly invented phonograph at the Paris Exposition, a little-known French writer named Villiers de l'Isle-Adam embarked on a novel in which Edison was to be the hero. The novel, whose working title was "Edison's Paradoxical Android," was not published until several years later (1886), and renamed *The Eve of the Future*. The story concerns the manufacture of the perfect woman. Through a secretive, newly developed form of dark magic—a combination of electricity and desire, photography and clairvoyance, magnetism and misanthropy—an ideal being, or "New Eve," is constructed.

Thomas Edison, Villiers was at pains to point out in his preface, was not really the hero of the book. The hero was Edison's legend, "the Magician of the Century, the Wizard of Menlo Park, the Father of the Phonograph." (A similar sentiment had been expressed by Mary Shelley, writing about Dr. Erasmus Darwin in her preface to the 1831 edition of *Frankenstein:* "I speak not of what the Doctor really did, or said that he did, but, as more to my purpose, of what was then spoken of as having been done by him.") Edison's legend, in other words, was already a fiction, available for use by anyone's imagination, even in the lifetime of its originator. Villiers justified his choice with a carefully chosen analogy: "If Doctor Johannes Faust had been a contemporary of Goethe and had given rise to his symbolic legend, would *Faust* not have been permissible nonetheless?"

In the book, Edison is portrayed as someone rather different from the plain-talking, hard-working ex–railroad telegrapher he appears to have been in life. In Villiers's version he is more like an alchemist, surrounded by puffs of blue smoke, disembodied voices, and mechanical birds of paradise: something more fantastical than electricity is at work here, and sets the scene for his secret project.

The story, then, is this: into Edison's world of magic and genius comes a friend, the young English nobleman, Lord Celian Ewald. Brief reference is made to an occasion on which Ewald saved Edison's life, and he arrives now in a desperate state, prompting Edison to vow to return the favour. Ewald's problem is that he is in love. He is in love with the outer shell of a beautiful woman—a perfect woman, Miss Alicia Clary. He likens her appearance to the famous statue of the Venus de Milo in the Louvre: she is, he says, "truly the splendour of the *venus victrix* humanised." But Alicia's soul is so disparate from her beauty that she seems "imprisoned, by some sort of occult punishment, in the perpetual denial of her ideal body." Her

132

voice is lovely, but what she says is naïve and vulgar; her eyes are wonderful, but when she looks at Ewald her gaze is empty, imperfect. The mismatch between body and soul is unbearable, and Ewald is at his wits' end: he has come armed with a revolver so that he can end his life, since her kiss, as he tells Edison, "only awakens in me the taste of suicide."

Edison, however, has a solution. Via a process of "transubstantiation," he will separate Alicia's body from her soul, and in twenty-one days' time, he tells Ewald, "at the same hour, in this very spot, Miss Alicia Clary will appear to you, not merely transfigured . . . but dressed in a kind of immortality. [She] will no longer be a woman, she will be an angel: she will be not a mistress but a lover; no longer Reality, but the Ideal."

He takes him to an underground laboratory outside "the world of the living," via a labyrinth illuminated by red and blue flashes of light. He calls out the name "Hadaly" (said to mean "ideal" in Persian) and "a kind of Being" emerges, surrounded by exotic birds that repeat by phonograph, as Edison explains, voices recorded from birds now dead. Villiers describes the figure as "a feminine armour, made of leaves of burnished silver of a matte and radiant white," which "revealed, moulded with a thousand perfect nuances, svelte and virginal forms." Hadaly, who wears a black mourning veil over her face (who, one wonders, has died?), is not yet a living thing. Edison explains that she does not contain a human being, she is only "a magnetoelectric entity . . . a possibility." She is waiting for the form in which she is to be incarnated, and waiting, more importantly, for someone to need her existence. Ewald's desire and desperation are what will finally bring her to life. Edison takes the postcard-size photograph of Alicia

Clary that Ewald has given him and projects it, through a series of complicated rays and reflections, on to a piece of white silk in a vast frame, so that her image is in colour and life-size. This, he tells Hadaly, is who she will become.

A large part of the book is taken up with descriptions of the android, long expositions of her internal make-up, as if Dr. Frankenstein were giving an anatomy lesson after the fact. Indeed, when Hadaly's "feminine armour" opens up, Edison comments that it is reminiscent of the work of the famous sixteenth-century Flemish anatomist Vesalius (who was one of the first dissectors of human bodies, and was sentenced to death for the practice). When Ewald complains that the minute detail with which Edison describes Hadaly's insides—the "objects of love"—is "truly hellish," he replies that it is only a problem if one is actually in love: "Do you think a doctor feels uneasy before a dissecting table, during an anatomy lesson?"

These mentions of anatomy are all the more disturbing for the fact that Hadaly's organs bear little resemblance to a human being. She is mechanical, electrical, magical. Her lungs—not unlike Kempelen's earlier drawings of lungs as bellows—are made of two golden phonographs, placed at an incline towards the middle of the chest, which pass metal sheets of recorded speech between them. There are twenty hours of "celestial chatter" in total, repeatable through steel styluses in an infinite number of combinations. Ewald comments that golden phonographs must be more beautiful than real ones, and Edison explains that he chose gold for two reasons: it doesn't rust, and it has "a more feminine resonance." In fact, "it's worth noting," he says, "gallantly," "that in composing a woman, I found myself forced to resort to the most rare and most precious substances, which in itself is a tribute to the enchanting sex. Nevertheless, I have had to use iron in the joints"—but then he adds that there is iron in all of us, in our blood—and to prevent rust, these joints are to be oiled with rose oil, which he takes from a shelf and gives to Ewald. There is enough in the bottle to last a century.

The android is threaded through with brilliant wires, which are "exact replicas of our nerves, our arteries and our veins," and she is set in motion by an "electro-magnetic motor," a spark that is said to be "the legacy of Prometheus." Ewald is worried that they are "tempting God" with this mechanical monster, but Edison has an answer to that too: "Since our gods and our hopes are no longer anything but scientific," he declares, "why should our loves not be scientific too? Instead of the Eve of the forgotten legend, the legend science finds suspect, I am giving you a Scientific Eve."

Days later, the New Eve is ready. "I should warn you," Edison tells Ewald, "it's even more frightening than I thought." But before he can meet her, Ewald is invited by Alicia for an evening walk around the garden. He is annoyed, impatient; he can think only of the perfect woman he will meet when he returns. Gradually, however, he warms to Alicia. He notices that this stay with Edison has made her quieter—there are none of her usual silly interjections—and she is more sympathetic to him. (Ironically, the father of the phonograph has taught her to shut up.) By the end of their walk he is wildly in love with her, and filled with self-hatred for being interested in "that dark prodigy of an Android." "Ah!" he murmurs, "was I insane?" He cries out to Alicia: "Oh, beloved one! I recognize you! You exist! You are flesh and blood, like me! I can feel your heart beating! Your eyes have shed tears! Your lips are as moved as mine are by a kiss! You are a woman whom love can make as ideal as your beauty!—Oh, dear Alicia! I love you! I . . ." But he doesn't finish his sentence. Alicia's calming hands are on his shoulders. "My friend," she says, "do you not recognize me? I am Hadaly."

Ewald feels duped: like Nathaniel in "The Sandman," he has fallen for the android, believing her to be more real, more "flesh and blood" than his lover Alicia; but Hadaly pleads—it is his desire that has brought her to life, she was made for him, tailored to his fantasies: he cannot leave her now, she will have no purpose. So despite his immediate sense of betrayal, he is persuaded, by her, and by Edison, to take Hadaly back to England with him. She is placed in an ebony coffin, constructed around her measurements. Ewald is to take her in the baggage compartment and animate her once they have arrived by touching one of the rings on her fingers. She is cut loose from Edison's laboratory; she is all his. They depart, aboard the transatlantic steamer, the *Wonderful*.

Some days later, Edison is surprised not to have heard anything from Ewald. Then, skimming through a newspaper, he finds this report: the *Wonderful* caught fire midway across the Atlantic, and a strange scene was caused by a young Lord E**** fighting his way through the flames to the baggage carriage. It took a number of men to stop him—there was no question of rescuing any luggage—and to get him into one of the lifeboats. Edison then scans down the list of fatalities and finds: Miss Emma-Alicia Clary. Not long afterwards, he receives a telegram from Ewald—perhaps a final suicide note—saying he is in mourning only for Hadaly, "that shadow."

Looking back on the book from this moment, death seems like a precondition for the entire story. If Ewald had not been prepared to die because of his lover's imperfection, Hadaly would not have been necessary. She is literally a life-saver, and can only be given life herself if she is needed to that degree. It is as if the whole extraordinary experiment of giving life to a machine could only be conducted in this particular limbo: the brink of death. It is notable, then, that Hadaly is dressed in mourning clothes, as is Any Sowana, the artist who sculpts her. Hadaly travels and sleeps in a black coffin, giving vampiric overtones to the tale (she feeds off life), and implying that in some way sleep and death must be the same. There is also a sense in which Edison's very project is an attempt to defy death, not just because Hadaly is not mortal (Ewald is instructed to destroy her when he is dying), but because in creating an unchanging being, Time, or the march towards death, is overcome.

This is inherent not only in the android but in the phonograph itself, as is clear from the actual responses to Edison's real invention. Recording a moment, as an image on paper or as a sound on metal, is a way of feeling one is stopping time, whilst time goes on regardless—which is why, as Roland Barthes has written, it "produces death while trying to preserve life." The photograph, he suggests—and, we might add, the phonograph—"mechanically repeats what could never be repeated existentially." In *The Eve of the Future* Edison tries to tempt Ewald by saying that "the dream of all human beings" is to "eternalise just one hour of love." The android, he promises, will be "nothing but the first hours of love immobilised." It will stop time—the time of Ewald's choice—and wouldn't

137

he rather "rehear" words that thrill him than be subject to the endless change of human feelings and speech? Like the real Edison's talking doll, set by a spring to repeat the same words over and over again, the android will stay the same for ever. Here Edison is proposing an end to depth and continuity: love as eternal return.

If death is the story's foundation, then its driving force is desire. After the android has been sculpted, Edison tells Ewald that it is like "the statue waiting for its creator Pygmalion"; but in that myth, Pygmalion himself is the sculptor—why would the already sculpted android have to wait? In Ted Hughes's translation from Ovid, "He dreamed/Unbrokenly awake as asleep/The perfect body of the perfect woman." Pygmalion sculpts her out of ivory, and, having fallen in love with his own creation, he prays to Venus for a woman like her. Venus responds by bringing Pygmalion's statue to life. Strictly then, Pygmalion is not the animator; he merely moulds the form (this corresponds, in Villiers's book, to Sowana's task); but since she is animated by the goddess of love, the animating force is love itself. Hadaly, according to this interpretation, would be waiting not for Pygmalion the sculptor, but for Pygmalion's desire. Without Ewald, Hadaly is only "a possibility"; without desire, the android cannot exist.

Iwan Bloch, a German sexologist and contemporary of Sigmund Freud, identified a "peculiar form of sexual aberration" known in the late nineteenth century as "pygmalionism." Pygmalionism, and its related tendency "Venus statuaria"—"the love for and sexual intercourse with statues and other representations of a human person"—were, in Bloch's view, closely allied to the phenomenon of necrophilia. He described

"Venus statuaria" first: "In the case of individuals who are sexually extremely excitable, a walk through a museum containing many statues may suffice to give rise to libido. Of this we have examples . . . as in the celebrated case of the gardener who attempted coitus with the statue of the Venus de Milo."

"Pygmalionism," on the other hand, involved women pretending to be statues. It was, Bloch wrote, "an imitation of the ancient legend of Pygmalion and Galatea, and utilization of this legend for erotic ends. Naked living women, in such cases, stand as 'statues' upon suitable pedestals, and are watched by the pygmalionist, whereupon they gradually come to life. The whole scene induces sexual enjoyment in the pygmalionist, who," Bloch added, "is generally an old, outworn debauchee."

Bloch ended his passage with a brief note on early sex toys, establishing a link between the android makers of the eighteenth century and future manufacturers of objects of desire. "In this connection," he concluded, "we may refer to fornicatory acts effected with artificial imitations of the human body, or of individual parts of that body. There exist true Vaucansons in this province of pornographic technology, clever mechanics who, from rubber and other plastic materials, prepare entire male or female bodies . . . Such artificial human beings are actually offered for sale in the catalogue of certain manufacturers of 'Parisian rubber articles.' "

In his essay on fetishism, Freud describes a fetish as the result of castration anxiety. He offers the example of a young boy, who thinks his mother has a penis. When the boy sees that she hasn't got one, he attempts to deny this fact, since, if she has been castrated, then he may be too. The fetish—desire transferred to another location, or a part desired instead of the

139

whole—is instrumental in the denial. The institution of a fetish, in other words, is a response to horror at the sight of the female genitals. It is perhaps no accident that Edison's real doll had a male torso: in Villiers's novel, the android and its accoutrements are spoken of in fetishistic terms, and at the West Orange laboratory extensive research was conducted so that the doll could be made from the most perfect pieces. The fictional Edison described his creation by breaking it down into parts, and the real Edison constructed his doll in a similar way, with none of the workers who made it seeing the whole, final result.

In Villiers's novel, Edison comes across something in his laboratory, and stays, staring at it, for some time. It is, we are told, "a human arm placed on a violet cushion." This extremity of a young woman is inanimate, but so lifelike it is said to be "as cruel as it is fantastic." It has a gauze bandage around its wrist, through which a small amount of blood can be seen. When Ewald arrives, Edison shows it to him. Ewald picks it up and notices that it is warm. He shakes the hand; it returns the pressure. He marvels at the "relic," and asks if it is not flesh. "Oh, it's better than that," replies Edison. "Flesh ages and fades; this is composed of exquisite substances, elaborated through chemistry, in such a way as to confound the adequacy of Nature." This fictional Edison is working on his android in pieces, and he even preserves some, disembodied, for admiration in his private office.

Throughout the book, the tiniest parts of Hadaly are detailed. Her feet, for example, are said to be made of silver. "A few samples of incomparable eyes" are delivered, in a box. An eminent dentist is called upon to take a cast of Alicia's teeth, mouth, and tongue, so that the exact shape can be transferred to the android. Eyebrows and eyelashes are fashioned from a lock of Alicia's own hair. The hair on Hadaly's head is the subject of a special, secret assignment: one of Edison's mechanics is sent to the best wig-maker in Washington. He takes a specimen of long, wavy dark hair, with a note indicating, in milligrammes and millimetres, the weight and length of the replica he'd like made. It is accompanied by four life-size photographs of a masked head, an exact mould of the skull, and some perfume—even Alicia's smell is reproduced.

There is an incident later on in which a group of spies, detectives, and gossip-mongers are trying to find out what Edison is up to, thinking he must be developing new electrical inventions in secret. They intercept a night-time delivery of a large box and open it. The spies are disappointed to find only:

> a brand new blue silk dress, some deliciously fine ladies' stockings, a box of perfumed gloves, an ebony fan encrusted with precious sculptures, some black lace trimmings, a lightweight and ravishing corset with bright red ribbons, fine linen dressing gowns, a jewellery box containing some quite beautiful diamond earrings, rings and a bracelet, bottles of perfume, a set of handkerchiefs embroidered with the initial H. and several other objects of that kind, in short, a complete womanly trousseau.

On seeing this, the spies think Edison has staged a joke on them, but what they have found is a bigger clue to the whole nefarious project than they could possibly realize. Given the way in which the perfect woman is viewed by these modern Pygmalions—a sum of manageable, impeccable, sexualized parts—the trousseau is, in a sense, more woman than the android. Corsets and jewellery, stockings and lace: why bother to animate her? What more could a fetishist want?

In writing the book, Villiers clearly had a strong sense of the history of artificial life. He mentions Faust a number of times, as if the whole story were a kind of flirtation with the Devil, and he takes an epigraph, early on, from Hoffmann's "The Sandman." At one point, Edison, explaining the word "Android" to Ewald, offers a potted, megalomaniacally disparaging history of our subject:

> Do you remember, my dear Lord, those mechanicians of bygone days who tried to forge human simulacra? Ha! Ha! Ha! . . . Those unfortunates, lacking in sufficient methods of execution, produced nothing but derisory monsters, Albertus Magnus, Vaucanson, Maelzel . . . etc, were barely makers of scarecrows. Their automata deserve to be shown in the most hideous waxworks . . . They are laughter and horror amalgamated in a grotesque solemnity . . . Yes, these were the first attempts of the Android-makers.

Later on, Ewald, worried he is out of his depth, says, "but my name is not Prometheus. It is Lord Celian Ewald." "Every man," replies Edison, "is called Prometheus without knowing it." All men, in other words, are capable of creation, and more precisely in this case: any man can create a woman.

In fact, though, Edison cannot create a woman on his own;

he requires the services of another woman, Any Sowana. When asked by Ewald why he embarked on this project, Edison tells a moral tale about a childhood friend, Edward Anderson, who fell in love (at a performance of *Faust*) with a young temptress, Evelyn Habal. Anderson was ruined by his desire for this false, artificial "child": he left his wife and children, he succumbed to opium addiction, he went bankrupt, and finally, after Evelyn left him, he committed suicide. At the end of the book it turns out that Sowana is Anderson's widow, whom Edison has rescued from the depressive state in which her husband's death left her; but she has not completely recovered: she suffers still from "one of the great neuroses recognized as incurable: Sleep." In her new life as a sleepwalker, she is also clairvoyant. It was Sowana who foresaw all the conversations Ewald would have with Hadaly and made Alicia record her part of them on to the sheets of metal. It was Sowana who first inhabited the Android, a figure "imbued with both our wills" (hers and Edison's)—as if they had invented a procreation of the mind, in which an effort of the will rather than the body could produce an "electrohuman creature." Sowana is a victim of "human magnetism," she is constantly mesmerized, almost magical: when she sculpts, she is said to look as though she were drawing with rays of light.

What is striking in this characterization is that Sowana's gift is repeatedly referred to as a "neurosis." Edison claims to have discovered through her that "nervous fluid" is akin to "electrical fluid" in its strength. So, despite her other-worldly powers, she is perfectly intelligible as an old-fashioned (or, in Villiers's time, newly diagnosed) hysteric.

The women in the book come as three types: the hysteric,

the child (rather strikingly, Edison refers to all of the women at some point as "children"), and the doll (when Ewald finally repents of the entire project he shouts to the android he thinks is Alicia: "I dreamed of sacrilege . . . of a toy . . . of an absurd doll without feelings!"). If these categories were interchangeable for the fictional Edison, it should perhaps not surprise us that Villiers's story had a factual counterpart. The perfect woman, in Edison's real factory, came in the form of a child-like doll.

It is not known if Villiers ever saw the toy phonograph Edison brought to the Paris Exposition in 1878, or if he had heard of Edison's plans to insert it into a doll. The dates certainly make it possible: Edison's early doll design, with the sound-funnel shooting out through the head, had been patented by then, and Villiers had no doubt read the story told to the *New York World* and reported in the *Figaro* of 4 April: "Now the lover," Edison joked, "while waiting for his sweetheart to finish her toilet, can place on a phonograph a sheet of the pretty things she has said to him before, and so occupy himself for a time with her counterfeit presentment."

Edison's most recent biographer, Paul Israel, confirms Villiers's fictional propositions by associating them with the fact that Edison became involved in the making of moving pictures—the practice of reproducing human beings mechanically. But surely the project most like the Eve of the future was Edison's talking doll—a doll, as Villiers's android was said to be; a child, as Villiers's women were labelled; a fetish-object in itself, a part for the whole, a doll for the woman. The celluloid ordered by Edison's experimenters now conjures up the method of "photosculpture" by which Hadaly is made; the dental plaster they needed is inevitably reminiscent of the way the android's teeth were made; and what was the gold lacquer if not a version of Hadaly's silver armour?

Since Villiers's book was not published until 1886, and the phonograph doll did not go into production until 1889, the year Villiers died, the fictional android would seem to be premonitory, but it was not overly so: Edison had been thinking along those lines for some time. His first doll patent was issued in February 1878, before Villiers began to write; he received his second patent for it in May 1880, and after that he only returned to the project when he moved into his new laboratory at West Orange, the year after *The Eve of the Future* was published in France. Had Villiers foreseen Edison's ambitions? Had he known about them already? Or had he, rather, suggested them to the master in writing?

It is impossible to tell; the only existing record of Edison's reaction to *The Eve of the Future* is a letter of 1910 from the Villiers de l'Isle-Adam "committee" in Paris, thanking the inventor for his donation of $25 towards a statue of Villiers. Given this gesture, and his taste for his own mythology, it's unlikely Edison would have been displeased by the book. At the time of the phonograph's invention, the sound of a "counterfeit" woman was a mere quip for Edison, which Villiers expanded to its most monstrous degree. In many ways, however, it was not so very far from a truth Villiers could not have known about—as we can see from Edison's thoughts about women more generally.

There is evidence that Edison privately thought of women as perfectible creatures, machines, or products. A year before

he married his second wife, Mina, he wrote in his diary of trying to put together a perfect woman:

> Thought of Mina, Daisy and Mamma G [the wife of
> one of his colleagues]. Put all 3 in my mental kaleido-
> scope to obtain a new combination à la Galton. Took
> Mina as a basis, and tried to improve her beauty by
> discarding and adding certain features borrowed from
> Daisy and Mamma G. A sort of Raphaelized beauty,
> got into it too deep, mind flew away and I went to
> sleep again.

That same evening, Edison made a second attempt to visualize the feminine ideal: "I will shut my eyes and imagine a ter-
raced abyss, each terrace occupied by a beautiful maiden. To
the first I will deliver my mind and they will pass it down to the
uttermost depths of silence and oblivion. Went to bed. Worked
my imagination for a supply of maidens. Only saw Mina,
Daisy and Mamma. Scheme busted—sleep."

Another note in his diary, which was not published until
1948, contains an idea reminiscent of the rose oil that Ewald is
instructed to apply to Hadaly's steel joints. After a walk
through the city, Edison writes: "Saw a woman get into a car
that was so tall and frightfully thin as well as dried up that my
mechanical mind at once conceived the idea that it would be
the proper thing to run a lancet into her arm and knee joints
and insert automatic self-feeding oil cups to diminish creaking
when she walked." Later in the book Edison complains that
there is not much chance of meeting a woman like the famous
French beauty Madame Récamier: "Nature seems to be run-

ning her factory on another style of goods nowadays"—
Nature runs a factory; women are goods.

Edison was, however, keen for women to become emanci-
pated in some ways, at least. He made his views clear in an article he wrote for *Good Housekeeping* in 1912, entitled "The
Woman of the Future" (Villiers, had he lived, would never
have believed the coincidence). Subheaded "A remarkable
prophecy by the great inventor," the article was mainly an
advertisement for electricity. Electricity, Edison claimed,
would soon be responsible for housework, and women would
be free to use their brains. The journal commissioned an illustration to accompany Edison's words: in it, "the Wizard Electricity," an elfin figure with a glowing light bulb on his head, is
busy cooking, cleaning, sewing, ironing, and even putting children to bed, whilst a dreamy-looking woman sits surrounded
by leather-bound books. So far, so progressive; but the freedom offered by these new inventions was introduced into a
world in which, Edison thought, women had not yet taken the
opportunity to think at all. Electricity, Edison wrote, "will
develop woman to that point where she can think straight.
Direct thought is not at present an attribute of femininity. In
this woman is now centuries, ages, even epochs behind man."

Indeed, the only female members of Edison's family who
ever helped him, very briefly, in his work, were immediately
given male nicknames. His daughter Marion, who as a ten-
year-old, after her mother's death, sometimes joined him in the
lab, was called "George," a name she continued to use for
years afterwards in her letters home to Edison. When Edison
remarried, in 1886, his wife Mina got into the habit of noting
down the results of his experiments. Her reward for this was

the name "Billy," which also stuck—letters from Edison in the 1890s still address her as "Darling Billy."

In "The Woman of the Future," Edison made clear that one of the improvements brought about by electricity was the possibility of breeding a race of mental prodigies: "The children of the future, the children of the exercised, developed man, and of the exercised, developed woman will be of mental power incredible to us today." These eugenicist aims were all in the interests of perfection: soon, Edison predicted, humans would take up all the available space in the world, and "the less of that space which is occupied by the unfit and the imperfect, certainly the better for the race." So, as a by-product of the inventions Edison introduced, imperfection would be eliminated, and perfect beings would populate the world. Edison would be the father of perfect humans.

In October 1887, Edison was approached by two men, William Jacques and Lowell Briggs, who had spent the past couple of years developing a talking doll, using the "graphophone" patented by Edison's competitors, Bell and Tainter. He immediately saw an opportunity to review his initial idea, and agreed to collaborate on the project, forming the Edison Phonograph Toy Manufacturing Company. In February 1888, he put Charles Batchelor, one of his most loyal experimenters, on to the project, to try to perfect the mechanism; but the other component parts of the doll were the subject of a very different kind of investigation.

A story is told about Zeuxis, a Greek artist of the fifth century B.C. Zeuxis was commissioned to paint a portrait of Helen of Troy for a temple at Croton. He asked the residents of that city to assemble five of their most beautiful women, so that he could choose one to model as Helen. When he saw the women, however, he thought no single one of them was worthy of the subject—each was lovely in some way, but none of them, he felt, could have started the Trojan War. So Zeuxis chose all five and, selecting the most perfect part of each, he composed a portrait of the most beautiful woman in history—a composite of many.

In nineteenth-century America, most dolls were composites, miniature Helens for their time. According to Miriam Formanek-Brunell, patents that were registered as being for whole dolls were in fact only for parts of dolls, particular portions or devices; it was usual for doll-manufacturers to make only a certain piece of a doll. As she points out, the products of these men "suggest that they felt more at ease with producing parts of women's bodies than with intact wholes." Thomas Edison worked the same way as everyone else—he made the phonograph that could fit into a doll, and he had designed a metal torso to house it; but he seems to have gone to great lengths to arrive at the right combination of other body parts—he didn't just make a deal with local suppliers; he searched far and wide.

The archives at the Edison National Historic Site, where the remains of the West Orange laboratory have been turned into a museum, contain a series of letters written to Edison from one A. B. Dick. They come from all over Europe—Paris, Nuremberg, Vienna, Berlin—and they mention other trips, to London, Prague, Italy, Russia. All of them are about the phonograph doll. What was this mysterious trip, one wonders on seeing these letters, not mentioned in any book about Edison? Who was A. B. Dick, and what kind of envoy was he?

In 1886, Edison had been approached by a man named Albert Blake Dick, who had until then been in the lumber business in Chicago, and had just acquired the rights to Edison's patented mimeograph, a kind of electric pen that could make copies as it wrote. Edison agreed to let him manufacture and market it as the "Edison Mimeograph," and the pair kept in touch (Alfred Tate mentions Dick in his memoirs as a fine example of a man with initiative). By 1889 Dick's new business was in full swing: he had a fancy letterhead with the initials "ABD" swirling into each other in the top left-hand corner, and boasted a trade fit for the most avid of clerks: "A. B. Dick Company," his stationery read, "Makers of labor-saving office devices, 152–154 Lake Street: The Edison Mimeograph, The Dick Adjustable Arm Rest, The Sherwood Cabinet Letter Files, Check Sorters, Order Holders, Time Files, etc."

How Dick became involved in Edison's next project is unclear, but he was evidently hired, early in 1889, to investigate the doll business in Europe. Distribution was not his only concern, however. Although he took one or two dolls and several doll phonographs with him to show to manufacturers there, the design was clearly not the final one, since, as is clear from one of Dick's letters, Edison was still planning to let the sound out through the head. The idea for the pepper-pot design in the doll's chest came, via one of these letters from Dick, from Emile Jumeau, the most successful doll-maker of the nineteenth century, and one of the most famous of all time. So if amendments were still being made to the doll—indeed, if it was still at that very early stage at the time of Dick's tour—then his purpose cannot simply have been to pave the way for European trade. No, Dick went to find out where the best dolls, or best doll parts were made. He travelled for months around the continent, sending missives full of optimism at times and disappointment at others. Just as Villiers de l'Isle-Adam had described the anatomy of his fictional android organ by organ and limb by limb, just as Zeuxis had been forced to piece Helen together from the attributes of five different women, Dick was on a mission to bring back the ideal body parts for the ideal doll, wherever they were to be found. With this in mind, his letters read like a peculiar travelogue—a journey to the perfect woman.

"My Dear Mr. Edison," Dick writes on 24 May 1889, from the Hôtel Castiglione near the Tuileries Gardens,

I have just about concluded my investigation of the Doll trade in Paris, and contrary to my first impression, I have found it confined principally to one concern, doing a business of over 300,000 dolls annually.

There are twelve doll makers in this city, but excepting the business done by M. Jumeau above referred to, all are in my estimation doing a very light business. As a class they are not strong financially, and are to be found mostly in lofts located in cheap districts and, like many other businesses conducted here, bear the stamp of being run on "hand to mouth" principles.

Jumeau is represented as the only manufacturer in France who makes every part of the dolls he sells, from the wigs to the soles of the shoes, and his sales of dolls large enough to contain the Phono I have with me, and larger, amount to about 100,000 per annum outside of the United States. He is also strong financially, and has taken more interest in the Phono than all the others combined.

His factories are large, and he employs over 1000 hands, mostly women and girls.

All Jumeau's dolls are made with paper maché bodies, having moveable hands, arms, feet and legs, as well as moveable head[s], all on the ball joint principle, and it is the general opinion among doll dealers I have seen that this feature should not be discarded, but that the shape of the Phono should be changed if possible so as to conform to the present style of doll bodies. The paper maché bodies can as you know be molded in any shape, and with a door at the back, or so as to separate in the center—the question is can the Phono itself be properly fastened into these bodies? I think it can and have instructed Jumeau to send you several sizes of *bodies* only which you can experiment with if you so desire.

Sinister overtones impose themselves on the practical matters detailed in these letters: after reading Villiers's novel and Edison's diary, the man moulding the female body "in any shape" no longer seems an innocent doll-maker but has become an industrial Pygmalion. One pictures Edison receiving a shipment from Paris of bodies in different sizes, packed up together like the surrealist concoctions of Hans Bellmer—what kind of "experiment" did Dick have in mind?

These weren't the only dolls Edison received. Dick sent parts of dolls or whole ones, mostly "undressed," from each city where he had met with any success ("When I get to

Nuremberg I will send you some more dolls," he wrote); but he had trouble knowing exactly what Edison wanted. On 8 June he wrote from Vienna:

> Before I left home you requested me to send you samples of different sized dolls with prices in quantities for the U.S. trade. I find the line of dolls so extensive and prices so varying that I do not know what to send you. I can get dolls large enough for your Phono, with kid or paper maché bodies all finished, in the best styles (undressed) at from 18 to 22 marks per dozen, and the better grades run up to 30 to 40 marks per dozen. Would it not be the best plan for you to select such sizes and grades as you may determine upon at Macy's or Schwartz in New York, and when you are ready to buy, send them over to me and let me get special quotations?

Dick's was a pertinent query. It seems likely that even Edison did not know what he wanted. While Dick was on his travels around Europe, another piece of research was commissioned by Samuel Insull, Edison's former private secretary and the manager of his machine works. J. S. Kelly's report on the doll trade in New York came from in-house, from the "wire insulating department" of the Edison Machine Works, and his job was partly to find out just what Dick wanted to know: what was desired of a doll? Kelly reported that the doll would not be marketable if it departed too much from "the prevailing style," and he attached a detailed description, "gleaned from conversations with either the buyer or one of the principals of nearly every large house [store] in the City," of what was and was not acceptable in a doll. By extension, his careful anatomy serves as a somewhat refracted snapshot of the feminine ideal at that time.

"Dark-haired dolls are unpopular," Kelly wrote.

In general four-fifths of the dolls made have brown or decidedly blonde hair and the remaining fifth have black hair. "Human Eyes" should be used in preference to the cheap glass eye. The hair should be long and banged or curled in the front, and should fall freely from the head; braided hair or anything of that sort being decidedly out of style. The lips should be apart slightly showing the teeth.

The very large and clumsy doll is now very seldom called for; the most popular sizes running 18 in., 20 in. and 22 in. in height. An 18-in. doll with "human eyes," good bisc head, long flowing hair, patent-jointed body, etc. can be imported by any large importer at about $7.50 per dozen. Stirn & Lyon, of 20 Park Place, this City, kindly volunteered to furnish samples with the hope that when we are ready to import we may favor them with an order.

Yet more dolls, then, were streaming into Edison's laboratory. Kelly's wording is hauntingly like some of Villiers's, made plain. The "Human Eyes" are reminiscent of Hadaly's, brought to the underground laboratory in a box; the specifications about hair are like those given to the android's wig-maker in Washington; the patent joints sound as though they might well need rose oil administered to them. And there is something carnal, almost carnivorous, about a child-shaped doll unchangeably exposing her teeth.

Back in Europe, Dick was experiencing a number of difficulties. In his letters he describes one memorable scene in which the doll-makers to whom he is showing Edison's design do not speak any English; they have to hire an interpreter to understand what the doll is saying. If the unintelligible spoutings of the phono doll did not seem downright demonic to these people, the magic and wonder of the thing had certainly vanished. Dick, clearly tired of his showpiece falling on deaf ears, asked to be sent some doll phonographs with recordings in French and German. In other letters, he makes his mission sound almost like that of a spy, and describes the delicate manoeuvres he has had to make in order to proceed with his investigation. "The fact is these Frenchmen are very suspicious," he writes from Paris. "Consequently I have been compelled to work slowly, seek introductions thro' favorable channels to the parties I wished to derive benefits from, cultivate their acquaintance socially, etc., etc."

Soon, however, he was able to work more rapidly. From Paris he went to Nuremberg, where he was "greatly disappointed," since he had been informed "from various sources that the largest doll manufacturing business on the continent was conducted here, while I find it to be a jobbing center only and a place where dolls are dressed." From this dolls' dressing room he travelled about four hours north, to Sonneberg, where there were a great number of wholesale dealers. None of the factories were large, but he was surprised to find that even "150 to 200 hands" could, as he so charmingly put it, "grind out a very large number of dolls daily." He found this to be the cheapest place to manufacture dolls, but that most of the output was of a lower standard, and a smaller size, than Edison required. Dick did, however, add one important note from this city. It wasn't about something he'd found; it's just that Edison's interest in what he failed to find gives the tale another twist. "So far," Dick wrote in early June, "I have not been able to find anything in the way of large mechanical figures for you. Altho' I have found a figure maker at Sonneberg, I could not find anything in his stock of samples which would please you. I may succeed better here in Vienna." Did Edison, then, want his doll to do more than just talk? Was he planning for this miniature android to become even more lifelike? As if a voice was not enough, he wanted to animate his creation with movement as well.

Dick's journey appears to have ended when he was instructed to meet Alfred Tate at the Hotel Metropole in London around 10 July. Tate was in Europe in order to pave the way for Edison's arrival at the Paris Exposition in August. Edison's trip was, of course, a grand success. Everywhere he went, people followed him. He was called "His Majesty Edison," "Edison the Great." He became a true heir to Vaucanson when he exhibited the perfected phonograph to the Academy of Sciences and, Dick must have been relieved to know, 30,000 people a day came to hear twenty-five phonographs speaking in as many languages.

On Edison's return, doll production began in earnest. He was hoping to have dolls in the shops in time for Christmas, but there were too many hiccups, and the factory resumed work more calmly in January. After Dick's many suggestions,

Edison settled for the metal torso he had patented with the miniature phonograph, for American wooden limbs, American hands and feet made of papier mâché, and a porcelain head manufactured in Germany by a company named Simon & Halbig, though its features were modelled on the style made famous by Jumeau of Paris.

The perfect woman could never be exactly perfect, however. A large part of Albert Dick's letters was taken up with the problems he was having, or the suggestions that were made in relation to the talking doll. Although Dick claimed that "all look upon the speaking dolls as simply marvellous," very early on, in his second letter from Paris, he reported to Edison that

a few slight objections have been offered to the present style of Doll Phono which I enumerate for your consideration.

1st The weight—This I think can be reduced, as the frame seems a little heavier than need be and perhaps other parts can be lightened—it seems that all doll makers strive to make their dolls as light as possible.

2nd Its dimensions—If it could be reduced in size to say two thirds of its present dimensions . . . it could then be utilized in the popular sizes on which there is the largest trade.

3rd The style or shape of the body or tin case holding the Phono—I am aware that this case can be shaped a little differently and have so talked to the manufacturers but they all seem to want something which they can put *inside* of their regular bodies of larger sized dolls.

Then Dick developed some objections of his own: the doll was too fragile for his, or for that matter any, purposes. "All but two of my dolls are already out of repair," he wrote from Paris. "The reproducing needle point has become loosened from the glass diaphragm . . . I think I can fix them up however." A week later he mentioned this problem again:

By the way, all of the small reproducing needles (or whatever they are called) have finally become detached from the glass diaphragm, and yesterday I bought some porcelain glue and repaired two of them very successfully indeed. They are very much better now than when I first received them, giving forth *fully twice* the volume of sound, and *perfect* articulations. I have been wondering whether the glue I used has had anything to do with the noticeable improvement.

He received no response to his suggestions, however, and continued with his negative reports. From Vienna on 8 June 1889 he wrote that

Although every doll maker without a single exception was highly pleased with the general idea of a speaking doll, a good many criticisms were offered on the present construction of the Phono, and its liability to get out of order, etc. An almost universal objection was

made to the weight and dimensions . . . Is it not possible to reduce the size to conform to their ideas?

A few days later, Dick hears that Edison has already
started to produce the phonographs, and writes, slightly
despairingly, "I hope you will note the content of my recent
letters, touching dimensions, weight, etc. before running out a
large quantity." Finally, Dick sends a curt note, reminding
Edison of his competitors, and giving him a brief lesson in
human nature. Dick is clearly certain by now that Edison is
wilfully ignoring the basic facts of the nursery:

> Mr. Edison, Could not paper cylinders similar to those
> used by the Graphophone Co. be used in the Toy
> Phono instead of the solid wax phonograms I have
> with me? These seem to be very fragile and are, I find,
> easily broken in changing to and from the Phono.
>   I am confident that the more delicate parts now
> contained in the Phono must be removed and more
> substantial parts substituted, particularly for use in
> dolls, as they will be handled mostly by children, who
> are not as a rule very careful. Yours truly, ABD.

The problem was that Dick and the people he met in
Europe were not the only ones to complain. Later that year
Alfred Tate received a memo from someone in the Edison
Electric Light Company, saying: "The good doll sent me was
broken. The other two do not talk very well. Can a tip-top
one—like I heard at the Laboratory—not be fixed up for me?"
And the difficulties did not stop in-house. Even when the dolls

were shipped out for sale there were complaints. On 24 April 1890, when the dolls were already in the shops, a toy-seller in San Francisco wrote to Edison's company that "among the samples received two were broken and four utterly useless for talking purposes." The very next day, a more lengthy expression of discontent was issued from Boston: "Gentlemen," wrote Horace Partridge & Co.,

> We are having quite a number of your dolls returned to us and should think something was wrong. We have had five or six recently sent back, some on account of the works being loose inside, and others won't talk and one party from Salem sent one back stating that after using it for one hour it kept growing fainter until finally it could not be understood. We should like to see somebody at once regarding this matter.

The most extraordinary letter came from a Mrs. H. M. Francis, who eloquently expressed, in her lament, the confusion between the human and the mechanical, between child and doll. Mrs. Francis bought a talking doll as a present for a friend, as she explains in her letter to Alfred Tate, before Christmas 1890.

> Much to my annoyance, when the Infant was received by him, it failed to say "Now I lay me down to sleep, etc." which had been firmly and mechanically introduced into its system. Of course, this gave him great disappointment, and more so for the reason that he had

a little girl about the height and possibly the age of the aforesaid talking Doll, whom he thought might learn this handsome prayer by hearing the doll repeat it.

What I should like to do, would be to send this Doll to you and ask that you use your kind endeavors and have it placed in such hands as will teach it to speak as originally intended.

What is wonderful about Mrs. Francis's letter is the deluded faith in the "firm" and the "mechanical," the notion that an inanimate doll might have an age as well as a height, and the idea that, despite the obviously mechanical means by which the words were "introduced into its system," the doll might also be "taught" to speak. The very existence of Edison's talking doll muddled all these distinctions.

Edison himself admitted his creation was not perfect. Late in 1890 he designed an improved model, since, as he noted, "there are too many expensive parts on the present one." As late as February 1891, *Harper's Young People* reported that Edison was working on yet another version, because he had found "twenty-five different faults in the present doll"; but Edison's battle to produce his perfect woman does not appear to have gone on for much longer. The dolls did not sell well, and the company finally folded at the end of 1891.

Though it is not known how many dolls exactly were sold, there are orders in the files for very few. Presumably they were sent to retailers in bulk, but there is only a record of a few individual shippings. For example, a man who helped Dick on his trip around Europe, Mr. Guldmann, asks for a doll for "a young lady in whom I take a very great interest." Another was sent to a church fair in Springfield, Massachusetts, two or three were to be exhibited in Australia, one was for the Emperor of Austria, two for the wife of Edison's long-time financial backer Henry Villard, one for a six-year-old girl in New York, one for Henry Ford, and another for one of Joseph Pulitzer's children. The superintendent of the phonograph doll works, E. McGovern, notes that he shipped 283 dolls to Edison's New York office on 22 September 1890, and one more on 20 October. That only adds up to a maximum of 294 dolls; and if Edison had set up his factory so that 500 dolls could be made a day, then where were all the rest of them?

On a scrap of paper in the archives is a hasty, handwritten list. "No. of Dolls reported in Doll Building Dec 4/90," it is headed, as if there were some sort of surveillance or detection involved. The dolls, as it shows, were scattered all over the place:

Centre Soldering Room	2251
Side Soldering Room	646
Passage way	471
Mr. McGovern's Room	609
Doll Room Lakeside Ave. side	340
In Centre	617
Other Side	1937
	6871
Office end of Doll Room	686
	7557

There were 7,557 dolls just in that building at the end of 1890; no more than a few exist today (none of the leading auctioneers or doll hospitals in New York or London have ever heard of Edison's doll now). What happened to them? What happened to the 100,000 dolls per year? The answer remains a mystery, and, as mysteries do, it has generated a number of rumours. Edison is widely believed to have destroyed his faulty creatures himself. At one time, historians thought he might have buried the dolls in a spot near the water tower. The tower still stands, a tall, white balloon announcing "EDISON" in large black letters; but the site has been thoroughly inspected with metal detectors, and nothing—not a single armour-plated body—was found. So no one knows what became of these little talking Eves of the future, but they are all gone now—victims, we may presume, of their own imperfection.

CHAPTER FOUR

Magical Mysteries, Mechanical Dreams

Any sufficiently advanced technology is indistinguishable from magic.

—Arthur C. Clarke

On the south side of London's Piccadilly, facing the fashionable tailors of Bond Street and flanked by sombre Georgian
red-brick houses, there stood, for almost an entire century, a
towering Egyptian temple. It was a madman's construction,
often thought absurd or uncouth, and always thought unfor-
gettable. Above the two stone pillars at its entrance stood statues of the Egyptian gods Isis and Osiris. These, in turn, were
guarded by a pair of sphinxes and, a decade before the deci-
phering of the Rosetta stone, the façade was covered in mysterious hieroglyphics.

The Egyptian Hall, as it became known, was built in 1812
to house a private collection of curiosities. A stuffed giraffe, a
rhinoceros, an elephant, a lion, and 3,000 birds and fossils,
some of which had been brought back to England by Captain

Cook, shared space with African weapons, North American clothes, and strange *objets d'art*. In 1816 it was home to the most popular London exhibition of all time: Napoleon's carriage, captured on the evening of the Battle of Waterloo, was shown to 10,000 people a day. Alongside it were the emperor's surviving coachman, his horses, and his travelling case, which still contained, among other things, a bar of Windsor soap.

Over time, the rooms were rented out and filled with spectacles of various kinds: art galleries, freak shows, clairvoyants, and conjurors. Phineas Taylor Barnum brought his midget Tom Thumb here in 1844, and the French illusionist Henri Robin performed a trick in which he lifted a child in the air by a single hair on her head. There were moving panoramas, clever silhouettes, ventriloquists, and jugglers.

By the 1870s it had become "England's Home of Mystery," and famous mostly for its magic. It was run by John Nevil Maskelyne, a descendant of the eighteenth-century astronomer royal, and by his friend George Alfred Cooke, a cabinetmaker. Maskelyne started out as a clockmaker's apprentice in Cheltenham, and continued to construct automata. At the Egyptian Hall he exhibited "Psycho," the nearly life-size figure of a Turk (based on Kempelen's Automaton Chess Player), now held in the collection of the Museum of London. Psycho could play whist, and guessed words or numbers the audience wrote down, by selecting cards from slots in front of its hands. Another android, "Zoë," sat on a high stool and drew portraits of famous figures of the day—Charles Darwin, perhaps by way of a sly comment on our gradual evolution from man to machine, was among her subjects. Two musicians—a euphonium player and a cornettist—were mechanical doppelgängers, automata Maskelyne made to look exactly like himself.

Maskelyne and Cooke made their names as showmen when they exposed the fakery of a then-famous illusion—the magical cabinet of the Davenport Brothers, who claimed to be American spirit mediums. Throughout his life, Maskelyne continued to attack anyone who called himself a spirit medium, though to the casual observer, the line between Maskelyne and the necromancers may not have been all that clear.

Ever since the eighteenth-century Belgian optician Gaspard Robertson had invented his pre-cinematic spectacle, the "Fantasmagoria," in which ghosts seemed to appear out of smoke in the crypt of an abandoned Paris chapel, miraculous events of this nature had become a conjuror's staple. One popular trick involved a man appearing to dissolve into the form of a skeleton. Maskelyne himself regularly caused his wife to levitate, and he developed a signature trick of spinning more than thirty porcelain plates at a time. It wasn't the spiritualists' tricks he minded, only their claim that they were caused by the dead. Magic for Maskelyne was a question of skill. It was the province of the living; but the idiom of necromancy was something he was happy to share.

The exposure of charlatans, and the intended triumph, through popular spectacle, of reason over mystification, occurred repeatedly in the eighteenth and nineteenth centuries, at moments when there was some new cause for science and belief to oppose each other. While La Mettrie confounded the theologians with his mechanistic view of human life, the mathematics professor Henri Decremps wrote a handbook on magic, revealing its tricks at a time when all of Europe was in thrall to the necromantic claims of Cagliostro.

As the historian Grete de Francesco argues, however, these Enlightenment men could only transpose, not eliminate, belief in unearthly forces. The mythology of the eighteenth-century quacks "hid itself and became entrenched where the Enlightened least expected it," de Francesco writes, "behind modern science itself, behind technology. Science and technology became magical." The progress of science in the nineteenth century—the invention of the photograph, the phonograph, and other apparent resurrections—coincided with a new popularity in shows of clairvoyance or dark magic.

While Maskelyne was performing his ambiguous magic tricks in London, on the other side of the Atlantic Thomas Edison was busy animating "mere dead matter" with his phonograph. Within a decade of that invention, Edison was caught up in a race to pioneer another form of animation. His competitors came from the international worlds of science and magic—worlds whose boundaries would be endlessly blurred by this new art: the art of moving pictures.

By a curious irony, the Egyptian Hall was to become one inspiration for a form of entertainment that would cause stage magic to lose its widespread popularity for good. In the audience there in 1884 was a twenty-three-year-old Frenchman named Georges Méliès, who, twelve years after his regular visits to the Egyptian Hall, incorporated the tricks he saw into movies, and became the inventor of the fiction film.

Kempelen had called his Automaton Chess Player "an illusion." Since then, the nineteenth century had introduced illusions of another sort: the disembodied voices of the phonograph, "mirrors with memories," as the first photographs were called, and the eerie performances of the Egyptian Hall. The Uncanny, in other words, had left its physical, concrete self behind; it no longer solely took the form of a single automated figure, but had become generalized, diffused throughout a new world of spectacle and magic. Into this world came another mechanized monster: the celluloid frames of the cinema, edited together by technological Frankensteins and brought to life. On film, man was made mechanical, reproduced over and over like an object in a factory, and granted movement by the cranking of a machine.

Cinema was a direct descendant of the androids of the Enlightenment; its birth was a Promethean, or Pygmalion-esque, event. Many years after pictures came to life, Thomas Edison made the first film version of *Frankenstein*.

The year before Edison put his talking doll into production, he and his colleague W. K. L. Dickson attended a public lecture in West Orange, New Jersey. The man they had come to hear was Eadweard Muybridge, a British photographer who had been experimenting for more than a decade with sequence photography—taking consecutive pictures of people or animals in motion.

Muybridge's project began when he was recruited by the American industrialist Leland Stanford to help resolve a bet. There was some controversy at the time over whether a horse ever had all four feet off the ground when running, and Stanford, who owned a famously fast horse, wanted Muybridge to photograph it while it moved at full speed. At the side of a running track in California, Muybridge set up a long shed containing a row of cameras, laid out at intervals of 21 inches. He painted the wall behind the track white, and drew vertical lines along it, which were numbered at the top, so that movement over time could be traced as precisely as possible over space. The shutter of each camera was attached to a wire running from the shed to the wall, which would be tripped by the horse as it ran along the track. Finally, he managed to obtain a series of successive images of the horse, and Leland Stanford was able to declare that the animal did indeed, in the course of its gallop, have all four feet off the ground.

The inspiration for Muybridge's method came from various kinds of optical toy that were then very popular— zoetropes, praxinoscopes, and phenakistoscopes—all of which were composed, in varying degrees of sophistication, of a revolving drum that contained a series of drawings. They were the precursors to cartoons, and worked on the basis of "the persistence of vision," the scientific principle by which the human eye retains, for a brief and calculable period of time, the image it has just seen, though the object is no longer before it. In the early nineteenth century, scientists realized that if images were made to follow each other in quick enough succession, they would appear to be not still, but rather an uninterrupted moving continuum.

Muybridge followed up his animal research with a new toy of his own. The Zoopraxiscope, as he called it, was a glass disk on to which images copied from his photographs were painted, and which he would project in rotation on to a large screen (because of the distortion of the projector, the photographs themselves could not be used). It was described by the *Illustrated London News* as "a magic lantern run mad." Muybridge travelled the world giving lectures with his invention, and it was this Edison and Dickson came to see. Two days later, Muybridge and Edison met, and they discussed, as Muybridge later wrote, "the practicability of using the Zoopraxiscope in association with the phonograph." In October that year, Edison filed a "caveat" with the Patents Office—a document signalling that he was working on a given invention but had not yet perfected it. "I am experimenting," the caveat read, "upon an instrument that does for the Eye what the phonograph does for the Ear, which is the recording and reproduction of things in motion."

On Edison's return to West Orange a year later, after the Exposition Universelle in Paris, he found that a "photographic building" had been constructed in his absence, and that Dickson had been hard at work on Muybridge's proposal. Dickson had recorded a series of photographed images on to a glass disc, and linked it up to a phonograph. According to Dickson, he welcomed Edison back with a talking film of himself, doffing his hat and saying, "Good morning, Mr. Edison, glad to see you back; I hope you are satisfied with the Kinetophonograph." Then he counted to ten on his fingers, to show the synchronization of sound and image.

Edison, however, thought glass discs were not the way forward. In Paris, he had met the French physiologist Etienne-Jules Marey, who had been experimenting for some years with what he called "chronophotography"—sequential photography over time. Unlike Muybridge, who shot series of separate images, Marey had been recording the successive movements of his given subject on a single glass plate—the result was a beautiful ghostly sequence, as if the person in the picture were a spring, slowly being pulled apart across the frame—but when Edison met him, Marey had changed tack: he had moved on from glass plates and was using a new kind of film that had been invented by George Eastman that very year. In 1888 Eastman had brought out his "Kodak"—the first instamatic camera—which used film stock on paper strips. In 1889 he replaced the paper with flexible strips of celluloid. This was what Marey was using, and it was this that Edison chose for his experiments at West Orange.

For a time, however, other Promethean exploits required more urgent attention—Edison had to make sure that dolls could talk, for one—and this meant that it was two years before he and Dickson unveiled what they could have invented there and then: the Kinetograph moving picture camera, and its attendant peep-show projector, the Kinetoscope.

In 1894, the first Kinetoscope parlour opened to the public in New York. It showed, amongst other films, *The Sneeze*, the first known copyrighted moving picture. *The Sneeze* featured Fred Ott, an Edison employee, sniffing snuff and sneezing. Sneezing, as the Automaton Chess Player pamphleteer Philip Thicknesse had jokingly pointed out, was one thing machines could not do—and yet here was a mechanized image doing exactly that.

The Kinetoscope was a huge success, but each one, of which there were rows in the parlour, could still only be seen by one person at a time. A machine that could project films on to a screen, so that they could be seen collectively, had yet to be devised. The Kinetoscope was exhibited in London and in Paris that same year, and Europeans were delighted to discover that Edison had failed to copyright his invention outside

America. The race to invent the cinema, as a large-scale public spectacle, began.

Although the midwives of moving pictures were many, two men stand out as the direct descendants of the android-makers of the eighteenth century. One was a scientist, and the other was a magician.

Etienne-Jules Marey, a professor at the Collège de France, influenced historic combinations of men and machines in ways he could barely have intended. His photographs of birds in flight offered vital information to the Wright brothers when they constructed their aeroplane, and his interests and achievements perfectly complete the circle that La Mettrie and Vaucanson had begun to draw over a century earlier.

Marey's first publication was a paper on the circulation of the blood. He became the first person to prove how the elasticity of the arteries affected circulation—research uncannily related to Vaucanson's rubber man. His book *La Circulation du Sang* is illustrated with an engraving of a machine he had constructed, which reproduced "the different mechanical phenomena of circulation." The machine is a table surrounded by a system of tubes, levers, and pumps, set in motion by a wheel and a pulley-belt—it looks like a combination of a musical instrument and a toy train set. An inset shows a detail of the machine—the heart, in the form of a flexible sack attached to two large tubes or arteries and composed, at least in part, of rubber; but Marey did not give his machine a human form and, in explaining why, was insistent on the scientific, rather than entertainment, value of his research. "I did not," he wrote, "give in to the childish desire to make a sort of automaton."

What interested Marey most was movement that was invisible to the naked eye—either because it was under the skin (he was the first person to give an accurate graphic tracing of the heart's movements), or because it was too fast to be broken down into intelligible components. He constructed the sphygmograph, an instrument that measured a human pulse and transmitted its rhythm via wires on to a cylinder of paper, leaving a trace of ink—a process not unlike that of a modern ECG.

He began his studies of human and animal locomotion in 1870, before Muybridge had set up his apparatus at the running track. In order to conduct his investigations, he needed to be able to register many different kinds of information simultaneously: the relationship in time and space of individual body parts to each other, changing centres of gravity, the muscular power required to perform certain movements, and so on. A graph could not show all of these things; what he needed was a camera.

Like La Mettrie, Marey thought the body was an animate machine, governed by the laws of mechanics; he wanted to find a way of applying these laws to physiology, to record the movements of man in such a way as to make him intelligible as a machine. His photographic plates and strips of film show men broken down into mechanical movements: they are like time maps of the human body, the human machine. So Marey forms a crucial link between philosophers and scientists and those who were to be thought of as magicians, between those early mechanistic thinkers and the technological fantasists to come. Through him, the cinema can be seen as a technological relative of magical automata—another method of mechanically reproducing the mechanical in man. Etienne-Jules Marey confirms, to refract a phrase, the persistence of a certain vision.

Though Marey was filming scientific subjects on to celluloid strips at the same time as Edison was branding his own form of entertainment, for the Frenchman the only interest lay in analysing the movement registered on film, so projection at natural speed was as useless as looking at the subject directly. He wanted to see what the naked eye couldn't, and therefore the films he shot were most valuable to him when seen still, as a sequence of frames. This is how they are reproduced in books, so it is easy to forget that they are in fact movies, which, when projected, form extraordinary documents.

Many of Marey's films are preserved in the archives of the Cinémathèque Française, housed in the old military fort of Saint-Cyr. Some of them focus on particular body parts; they are untheatrical, yet beautifully lit; they portray a single movement, repeatedly, and then end. In one, a hand, framed at the wrist by the sleeve of a black jacket, is filmed against a black backdrop, so that the hand appears disembodied, like part of a classical sculpture come to life. It begins as a fist; it unclenches with a single, forceful motion, so that the finger muscles are spread out and taut; the fingers return to the centre, becoming a fist once again; they repeat the motion several times and stop. When seen frame by frame, halfway between the fist and the extended hand is a hovering, grasping extremity, which looks as though it is slipping from something, or threatening like claws. Detached from its experimental context, the hand seems to be borrowed from a horror film, testing a newfound movement, as if it were a monster trying out humanity, limb by limb and muscle by muscle—and in a sense, this is exactly what was going on, since animation through moving pictures was to bring with it a whole new range of monstrous or magical possibilities.

The hand in the film was not, however, some dug-up limb or automated sculpture but a hand attached to a real person. It belonged to a magician, Monsieur Arnould, or possibly Monsieur Raynaly. Both were well-known illusionists, who performed at the Robert-Houdin Theatre in Paris. They were hired by Marey and his assistant Georges Demenÿ for the purposes of an investigation conducted by Alfred Binet, the man who wrote about blindfold chess players. Binet was preparing a study of "the psychology of prestidigitation," and he invited the two magicians to his lab, where they showed him tricks, and broke them down into individual movements. When they were filmed, the components of the sleight of hand could be clearly seen.

Binet had already spoken to their employer, Georges Méliès, the magician who owned the theatre, but although Méliès was happy to answer a few questions, he refused to be filmed at work and sent Arnould and Raynaly in his stead. He could not give away his secrets, he explained to Binet—and his refusal was based on a crucial assumption: that the camera could never lie, and his tricks would be unmasked. This was the closest Méliès and Marey came to meeting. It was 1892: before the Kinetoscope arrived in France, before the Lumière brothers' Cinematograph was invented.

Yet four years later Méliès was to change film history for ever. If Marey was the cinema's strongest link to androids in terms of its form—he used a technological invention to show man as a machine—then Méliès revived that Promethean spirit with the content of his movies. By repeatedly filming stories of dolls coming to life, by endlessly reproducing mechanical tricks, Méliès transferred the quest of earlier android-makers to a new virtual reality. He made the human body do impossible things, and proved how mechanical or puppet-like our celluloid selves could be. Méliès, like Vaucanson, Kempelen, and Edison before him, tested the boundaries of what was human. By inventing trick photography in motion pictures, Méliès ensured that, in his own movies at least, the camera would never tell the truth.

Méliès was the son of a shoemaker. He was sent to London in 1884 by his parents, to work for a family friend who had a shoe shop in Piccadilly. It was hoped that Méliès would learn some English, but while he spoke none he kept himself entertained with the speechless wonders of the Egyptian Hall.

When he returned to Paris, Méliès took a job in the family business, overseeing the machines in the factory; but what he had seen at the Egyptian Hall would not leave him, and he began to frequent a shop in the Marais run by Emile Voisin, a supplier of conjuring materials. Voisin gave Méliès lessons in magic, and later that year he arranged for Méliès to give his first public performances at the Galerie Vivienne, which specialized in puppet theatre, and at the Musée Grévin.

The Musée Grévin was (and still is) a waxwork museum, less grand and more carnivalesque than Madame Tussaud's. There was a fast turnover of exhibitions in the nineteenth century, since it was intended to be a "newspaper in three dimensions," a series of reconstructions of recent crimes and other newsworthy events. The Grévin exhibited scientific wonders as well, however—it introduced its audiences to the phonograph, the telephone, the X-ray—and there was also an "optical theatre," in which Emile Reynaud, inventor of the Praxinoscope and a precursor to cartoon animators, showed magic-lantern slides. In another part of the museum there was a "cabinet fantastique," where Méliès performed his magic tricks. The scientific inventions and the magic-lantern slides were, for their audiences, new forms of life, housed in a museum devoted mainly to deathly simulacra; and Méliès, who shared space from the very beginning of his career with wax figures and marionettes, was one of the few animated exhibits in otherwise inanimate settings. He went by the name of Doctor Mélius: it was as if, by magic, or perhaps by medicine, he meant to bring what surrounded him to life.

Three years later, Méliès's father retired, leaving the family business to his three sons. Méliès sold his share to his elder brothers, making enough money to take up an opportunity that seemed to have been dropped by fate into his lap. Not far from his home, and up for sale that year, was the rundown magic theatre of the late Jean-Eugène Robert-Houdin.

As a child, Méliès had attended performances at the Robert-Houdin Theatre and, three years before his trip to London, he had done his military service in Blois, Robert-Houdin's home town, where the magician's estate had been turned into a museum. In Robert-Houdin's time, the estate had been the first electrically operated house in the world. The "magician of science," as one of his descendants has called him, took out a number of patents, right up until the last years of his life, on various electrical clocks and alarms. His home contained an automatic feeding machine for his horse, a burglar alarm, and a special device by which the great illusionist could put all the clocks in the house backwards or forwards, depending on when he wanted to have his lunch.

Robert-Houdin, the man known as "the father of modern magic," was not only an illusionist by profession but a myth-maker as well—and no myth seems more fanciful in his telling than his own. The person who would later claim in his mem-oirs to have repaired Vaucanson's duck, and who wrote an improbable embellishment of the history of Kempelen's Chess Player, started out as a watchmaker's apprentice. One day, he asked a bookseller for a book on horology and was given by mistake an eighteenth-century encyclopaedia of magic. He stayed up all night, he said, transfixed by the tricks in his new book.

Some time later, while working in another part of the Loire valley, Robert-Houdin fell victim to a serious bout of food poisoning. It was thought that he might die, and so, want-ing to bid a final farewell to his family, the young man ran away from his doctors and set off for Blois. On the way there, however, he jumped from the carriage in his delirium, taking a potentially fatal leap into the road. The hallucinations that ensued are described in his memoirs as a "fantasmagoria": he imagined that his head could be opened like a snuffbox, and he was reduced, he wrote, "to a mechanical existence."

On waking, Robert-Houdin found that he had been res-cued by a man who had not only trained as a doctor, but hap-pened to be a travelling conjuror as well. The conjuror, who called himself Torrini, nursed him back to health and began to teach him the art of magic, of which the young man was already an amateur. In exchange Robert-Houdin was able to bring his mechanical skills to bear on a machine of Torrini's that was out of order. The machine turned out to be an android, an automated harlequin, and from this moment on, Robert-Houdin saw that mechanism and sleight of hand could share the same world, be part of the same spectacle. He repaired the automaton, learned enough magic to give performances of his own, and finally returned to his family.

Robert-Houdin set up shop in Paris, where he sold timepieces and astronomical clocks along with mechanical toys. He decided, since he had learnt all he could of magic, that he would turn his attention to "the science, or rather art, of making automata." He attempted to study their history, and found stories of the automated Roman fly that acted as a deterrent to other flies, and kept the city of Naples free from disease. He read about Roger Bacon and Thomas Bungy, two thirteenth-century Franciscan monks who were said to have been instructed by the devil to make a brazen head, which would eventually acquire the power of speech and become an oracle. The monks spent seven years on their creation, and afterwards fell asleep, so that when it spoke they failed to hear it. Robert-Houdin learned about Albertus Magnus's chattering android, and he even found a record of the first blood machine ever made, the precursor to Vaucanson's mentioned in the *Journal des Savants* in 1677.

Nevertheless, Robert-Houdin remained disappointed. "Although I continued my inquiries," he wrote, "I only attained the unsatisfactory result that nothing serious had been written on the subject of automata.

" 'What!' I said to myself, 'can it be possible that the marvellous science which raised Vaucanson's name so high—the science whose ingenious combinations can animate inert matter, and impart to it a species of existence—is the only one without its archives?' "

Many of the automata Robert-Houdin made during this period became central to his magical repertoire: his "Pastry Cook of the Palais Royal," an automaton chef who would fetch real pastries from a miniature shop; his "Fantastic Orange Tree," which could bear fruit instantly and produce from its uppermost orange a pair of butterflies holding up a spectator's handkerchief; a tiny French guardsman who shot rings from the barrel of his gun on to a glove inflated on top of a glass vase; and "Diavolo Antonio," a mechanical trapeze artist the size of an infant. But perhaps his greatest mechanical feat was his writing automaton, which won a medal at the Paris Exposition of 1844 (the occasion on which he claims to have repaired Vaucanson's duck).

Robert-Houdin exiled himself to the suburbs of Paris in order to concentrate on his masterpiece. He asked two different sculptors to make a head for his automaton, which was dressed in Louis XV costume, almost certainly as an homage to the work of the mechanicians that king admired; but the first head was built to look like that of a saint, and the second had the blankness of an artist's mannequin. Frustrated, the conjuror decided to take on the task himself. He began to mould a piece of wax, and, as he tells it, quite by chance, the head began to resemble his own. He turned the android into his double and, in answer to the question "Who gave you life?," the automaton was able to sign, in an exact copy of his handwriting, Robert-Houdin's name.

The King, Louis Philippe, was apparently delighted with the android when it was shown, and after the Exposition the machine was bought by a travelling showman who had by then become famous. Phineas Taylor Barnum met Robert-Houdin, and attended his performances, while on tour with Tom Thumb in 1844. He saw the writing automaton at the Exposition and when the show was over, he bought it, as he wrote, "for a good round price." He exhibited the automaton in Britain and then shipped it back to America, where it stood in Barnum's museum alongside waxworks, skeletons, giants, and midgets.

Robert-Houdin frequently used his sons Emile and Eugène as collaborators in his magic tricks. After his death in 1871, the elder son, Emile, decided to take on the magical mantle, and for many years he continued to manage the Robert-Houdin Theatre. It was Emile's widow who decided to sell it, along with Robert-Houdin's original automata, to Georges Méliès in 1888.

Méliès immediately went about redecorating the theatre, and repairing the machines. He set up a workshop in his family's factory, where Eugène Calmels, a mechanic he had met while overseeing the machine works there, took time out from making shoes in order to help him reconstruct Diavolo Antonio, the Fantastic Orange Tree, the French Guardsman, and the Pastry Cook of the Palais Royal.

Méliès hired magicians from other theatres—Arnould, Raynaly, and Joseph Buatier de Kolta, whose "Vanishing Lady" trick—the most widely imitated stage illusion of the era—he had seen at the Egyptian Hall. He had an assistant named Marius, who was always willing to be decapitated in the interests of illusion, and a female star, Jehanne D'Alcy, whose pretty and petite figure made her perfect for disappearing acts.

He performed tricks that owed a good deal to Robert-Houdin, and even more to those he had seen at the Egyptian Hall. There was a Davenport Brothers spoof, a body chasing after its head, and an animated skeleton. Ever curious about artificial animation, Méliès devised a version of Hoffmann's "The Sandman" entitled "Coppelius's Dream."

He would often end a performance with a magic-lantern show, in which he used mechanised plates: by means of levers and ratchets, he managed to give impressions of lightning, snow, and moving trains. Between magic shows, people would wander through the foyer, where "Cagliostro's Mirror" would return their reflection as a bouquet of flowers, or the face of a beautiful woman. And there they would see, repaired and set in motion, the uncanny automata of Robert-Houdin.

When Edison's Kinetoscope came to Paris, Antoine Lumière saw it and devised a plan. He rented a workshop on a floor directly above the Robert-Houdin Theatre and, with his sons Louis and Auguste, began working on a claw-like action (similar to that used in a sewing machine to advance cloth) that would hold Kinetograph film steady when projected on to a large screen. On 28 December 1895, the Lumières gave the first performance to a paying audience of their newly developed Cinematograph, at 14 Boulevard des Capucines. There were a hundred seats, and the entry fee was one franc. The assembled audience experienced what was probably the most extraordinary collective sense of the Uncanny for centuries. One of the films was of a train arriving in a station; photographic images they had only ever seen as stationary suddenly came to life, and legend has it that as the train on the screen came towards them, several people, fearing an accident, ran from the room.

Méliès was there that day, and he immediately asked Antoine Lumière, his neighbour, if he could buy a Cinematograph. But Lumière refused, saying that the invention had been created for the purposes of science, not magic.

Though the Lumières had turned him down, Méliès was not to be thwarted. He heard about another machine, invented by Robert William Paul, a maker of electrical and optical instruments in London. Paul gave the first public showing of his "Animatograph" in January 1896. In February, Méliès went to London to buy one. He returned with a projector, some of Paul's own short films, and some Edison reels—seascapes, dancers, and boxing kangaroos. (David Devant, Maskelyne's new business partner, also bought an Animatograph for the Egyptian Hall.) When Méliès projected the films for the first time at his theatre two months later, one of the magicians there claimed he had seen the ghost of Robert-Houdin in the audience.

In the meantime, Méliès set up a workshop in the room at the theatre usually reserved for repairing the automata. With the help of two mechanicians, Lucien Korsten and Lucien Reulos, Méliès constructed a number of machines before he managed to make a camera based on Paul's projector, using spare parts from the magic stores of the Robert-Houdin. In Méliès's workshop, you might say, automata gave birth to the movies.

Eventually, in May or June that year, Méliès was ready to make his own films. In the garden of his family's country house in Montreuil, he filmed three people playing a game of cards. His cameraman, Leclerc, was the pianist at the theatre. Towards the end of that year, he started building a studio in the garden at Montreuil. It was an enormous conservatory, the first permanent daylight film studio in the world, and the stage was built to the exact dimensions of the Robert-Houdin Theatre, with a pit three metres deep in the ground to allow for the same trapdoors and ramps.

This was what separated Méliès from the Lumières: while the brothers were interested in making documentaries, for Méliès the possibilities offered up by the new medium were connected with conjuring. He immediately saw it as an extension of what he did at the theatre: not just another trick to his bag, but a way of bringing illusions to life in a way that had previously been impossible. He spoke of himself as a maker of "artificially arranged scenes": he was the first to realize the cinema's narrative potential, and consequently became the inventor of film as we now mostly know it—the inventor of film as fiction.

The film historian David Robinson has suggested that Méliès was inspired in his filmic trickery by a book of which he "would certainly have been one of the first purchasers." *Magic—Stage Illusions and Scientific Diversions,* by Albert Hopkins, was published in New York in 1897. It contained some old favourites with which Méliès was already quite familiar—an illustration of the Vanishing Lady trick, an illusion once performed by Robert-Houdin, and Marey's chronophotographs. The book also offered some tips on trick photography—how to photograph spirits, for example, and (something

Méliès would draw on later) how to photograph a human head, chopped off and placed on a table.

If Méliès really did own a copy of Hopkins's book, it would have been, in the strangely zigzagging history of mechanical life, a truly fortuitous possession, confirming the continuity between the androids of the past and the celluloid Eves of the future; because in amongst all that sleight of hand was a detailed account of Edison's talking doll. Long before the doll had become a footnote in the history of Edison's achievements, Méliès would have seen it, engraved as it had been on the cover of *Scientific American* seven years earlier. Hopkins was the editor of that journal, and it was he who had written the story: he was an expert on magic, overseeing the history of science.

Gradually, Méliès expanded his studio building. He added two extensions to the left and right of the stage, which functioned as wings and gave the stage area extra space, and he built two iron bridges for the camera operators to move around. He went on to add an elevator crane, which allowed people and objects to float in the air, and several extra buildings. There was a two-storey shed, which provided dressing rooms for men and women, and an enormous hangar for storing his hand-painted sets and bulky props (whole trains, boats, aerial balloons, and stairways were kept on stand-by). A fireproof warehouse housed over 20,000 costumes from all different periods, including wigs, shoes, jewellery, and weapons.

Méliès started out using his family and friends as actors. His neighbours, his servants, and his gardener were all recruited, and when extras were required they were culled from the factories nearby. Theatrical performers, he later wrote, thought it was beneath them to take part in this inferior medium; but Méliès found that the ordinary workers carried off their period costumes rather badly and were, he said, "lacking in chic." So he hired some of the dancers at the Chatelet Theatre, who were only too pleased to earn the extra pay. Soon the dancers from the Paris Opera joined in, then music-hall performers, and eventually, theatre actors as well. Méliès paid them one gold Louis a day, and served them lunch. But for many of his films, only Méliès himself could be relied upon to play the main part, so complicated had his technique become.

A famous, and possibly apocryphal, story is told of Méliès filming a street scene at the Place de l'Opéra. Halfway through, his camera jammed, and by the time he had fixed it, the scene had changed. He finished filming that reel, and when he developed it, he found that, from one frame to the next, men had turned into women and an omnibus had changed into a hearse. With this, he discovered the principle of the stop-substitution trick, and it was to become his signature.

Méliès usually woke up at six in the morning and was at the Montreuil studio, an hour away, by seven. He built and painted all his own sets, working until five, when he returned to Paris. He arranged meetings for the hour between six and seven, quickly had dinner, and was at the theatre by eight. During the show, he would draw sketches of his sets and story-boards in pen and ink. After the performance, he went back to Montreuil and continued working until midnight. Generally, he would spend the weekends shooting the films for which he had prepared the sets during the week. On days he was filming, the light would last on the studio stage from about eleven a.m. to three in the afternoon. All of the films were developed in Paris, at the Robert-Houdin Theatre. It was, he wrote, "a feverish life, with no respite, which was to last twenty years."

Méliès filmed magic tricks and fairy tales, a trip to the moon and a tunnel across the Channel; his was a world of dreams, or hallucinations. He became known as "the king of fantasmagoria," "the magician of the screen," the "Jules Verne of film." Though he also made a few documentaries, and "reconstructed newsreels" of the Coronation of Edward VII and the Dreyfus affair, he chose his fantastical subjects because they formed "an inexhaustible mine." "This new genre," Méliès wrote in his memoirs (which were written in the third person), "allowed him imaginative compositions, the most comical episodes, and, at the same time, the realization of things thought to be impossible. So in that genre, he found material which would satisfy primitives, but also intrigue scientists and give pleasure to artists."

By 1897 the Robert-Houdin Theatre was given over almost entirely to cinema, with conjuring tricks reserved for matinees. Between 1896 and 1912 Méliès made 500 films. Less than a quarter of them have survived.

In a set painted to look like Ancient Greece, a sculptor chips away at the statue of a woman. She is not a real woman, not even a real statue; she is, in the manner of primitive theatricals, a painted piece of cut-out board—she is doubly unreal. When the sculptor has finished, he seems disappointed. He gazes at his handiwork and bends down on his knees, praying to the gods, or wooing the statue. Suddenly, the board comes to life:

in its place is a woman, dressed in similar clothes and standing in an identical pose. Apparently unaware of the magic that has just been effected, the woman casually turns to look at the sculptor and, as if dissatisfied, descends from her plinth. She walks towards another plinth, which is further in the foreground, and stands on that one instead. The sculptor is amazed; but the woman, who remains indifferent and rather snooty, behaves as though nothing in the world could be more ordinary. She strikes a classical pose, as if pouring something from an urn into a chalice. The sculptor approaches; he walks around her, astonished; but as he gets closer the woman performs another conjuring trick: her urn and chalice turn into a harp. Now desperate to possess this changeable creature, the sculptor tries to grab her, and throws his arms around her legs; but he finds himself hugging thin air, and falls flat on the plinth: the animated statue has vanished, and reappeared on the other side of the room.

Second by second, and more and more manically, the sculptor and his former creation play a game of cat and mouse around his studio. As he puts his arms around her waist, she splits in two, leaving a skirt in the foreground and making her bust reappear, like a new sculpture, on a plinth just behind. The sculptor tries to catch the skirt and match the two halves, but the skirt runs away and the bust switches to another plinth. As he tries to grab the bust it reunites, of its own accord, with the skirt. Through all of this the woman has been calmly observing her creator spin into a comic frenzy, but now, whole again, she turns to give him a mocking, or perhaps a threatening, smile. She gestures as if to say, "See—you can't catch me!" and takes a sneering bow, her posture a vengeful travesty of a magician's curtain call. Then, regaining her former composure, she walks back to her original plinth, and turns to painted stone.

Like its subject, Méliès's trick film version of *Pygmalion and Galatea* disappeared, for almost a century, and reappeared in Spain in 1994. It is still listed as missing or destroyed in every Méliès filmography. The rediscovered print was handed over to the Centre National de la Cinématographie, the French state film archives, and restored there only recently. The CNC archives, like those of the privately funded Cinémathèque, are housed in an old military fort on the outskirts of Paris—as if the films needed to be preserved in solid secrecy. While I watch Méliès's *Pygmalion* there, the frames flickering across the small screen of an editing table, an archivist says he is surprised I have come all this way to see a mere sixty seconds of silent celluloid. But it is more than worth it: because although *Pygmalion* was produced in 1898, I realize that what I am looking at now should, if there were such a thing as cinematic justice, have been the first film ever made.

The Pygmalion myth is not only the subject of this individual film; it is also the story of the invention of cinema itself: what was once still comes to life. In his version of *Pygmalion*, Méliès, a magician familiar with the uncanny wonder of mechanical automata, has constructed a perfect metaphor for the magic of moving film. Everything that was wrapped up in the medium's early days is there: the desires, the fears, the superstitions, the power, and the hysterical zaniness of its first jagged steps.

Perhaps even more importantly, unlike the Lumières' *Train Arriving in a Station* or *Workers Leaving the Factory,* Méliès's *Pygmalion* was a trick—a trick played not only by the new medium on its audiences, but also by Galatea on Pygmalion. The magic doesn't appear to have been effected by the sculptor himself (played by Méliès, and therefore viewable as a magician figure). In Ovid's telling, it is Venus who brings the statue to life because Pygmalion has prayed for a woman like her: the animation is a gift; but here, the woman (Venus, if she has animated the statue, or Galatea, if she brings herself to life) is clearly not acting to please the man. Her mission is to tease, to parade her power, to look down on the mortal sculptor, and finally—unlike the ending in Ovid—to revert to an inanimate condition.

The film is the story of a dream (a beautiful statue comes to life) or a nightmare (a woman is sent to mock). It is both the story of Pygmalion and the story of Pandora, in which woman is sculpted by the gods and sent down to men as a punishment for Prometheus's transgression—the transgression of animating man. Either way, what the film represents, with all the anarchic freedom Méliès made his own, is the cinema as a living, moving impossibility. *The Voyage Through the Impossible* was another of Méliès's titles, one that could just as well describe the entire project that became his life.

Though Méliès is now mostly known for *A Trip to the Moon*, which he made in 1902, and for the fairy-tale scenes that led him to be admired by the surrealists, he also made a number of films that can be interpreted as comments on the very activity in which he was involved. Some are no longer extant, and only their titles show that Méliès was rehearsing an age-old concern—*Gugusse and the Automaton, Coppélia or the Animated Doll*. Others can be seen: Méliès brings mannequins, statues, and playing cards to life; he produces women from thin air and makes them vanish just as fast. He acts out, live on film, the very fictional tropes Freud listed as uncanny: he decapitates himself and plays with his head, he makes his own double appear out of nowhere; limbs are dismembered and inanimate objects given a life of their own.

These films, which might be called magical comedies, are ideal embodiments of the elements said by the philosopher Henri Bergson to be conducive to laughter. Bergson, a contemporary of Méliès, wrote that what was required in comedy was "a visibly mechanical articulation of human events." He spoke of "a kind of automatism": what he described as comic, Freud called uncanny and Heinrich von Kleist saw as grace. Kleist, writing about marionette theatre around the time Hoffmann wrote "The Sandman," tells of a professional dancer who aspired to the grace of a puppet, to have his limbs move like pendulums and his performance "transferred entirely to the realm of mechanical forces." "Only a god," the dancer told Kleist, "can equal inanimate matter in this respect."

If the cinema, for Méliès, was an extension of the automata he repaired and set in motion in his foyer, then people on film could be seen as androids too—mechanized men, distributed into tiny frames of celluloid, their movements broken down into mechanical functions. Méliès, now long dead, is visible in the films he left behind; but he is only the mechanical reproduction of a man—he comes to us via the filter of a flickering machine, an automaton in several dimensions.

The first trick film ever made was Méliès's *Escamotage d'une dame chez Robert-Houdin,* translated as *The Vanishing Lady.* "*Escamoter*" is an old conjuring word: it means to make something disappear by sleight of hand, and in this case, to

12/14/2007

make a lady vanish. But an "*escamoteur*" is also a pickpocket, a person with light fingers; by extension, the word has become associated not just with magic but also with less entertaining forms of fraud. It means to hide or erase something and, more figuratively, to get out of doing a chore, to slip through the net, or to make time fly.

So *The Vanishing Lady,* already the most popular illusion of the stage, was a complicated, double-edged affair—a trick of all tricks, in which the intention of the magician could be rather blurred.

The setting is a painted stage, decorated in ornate Louis XV style, like that of the Robert-Houdin Theatre. Méliès, dressed in evening clothes (his conjuring attire), enters and takes a bow. He brings a woman (Jehanne D'Alcy) onstage from the wings. Making her stand to one side, he places a sheet of newspaper on the floor, and puts a chair on top of that. He gestures to the woman to sit down, which she does. He takes a cloth from a side table and covers the woman with it. He swishes his hands in the air, takes the cloth away, and the woman is gone. The trick is Buatier de Kolta's, move for move.

But then Méliès swishes his arms in the air again, looking upwards, as if something is about to fall from the sky. All of a sudden, a skeleton appears on the chair. Lucy Fischer, a film historian who has put forward a convincing feminist analysis of these early films, describes the action that has just taken place as Méliès turning a woman into a skeleton; but this is not exactly what he does. He makes the woman disappear, and a skeleton arrives in her place, after a brief period in which the chair is empty.

Moreover, as John Frazer points out in his book on Méliès, *Artificially Arranged Scenes,* the second part of the trick is crucially different from the first. Until the appearance of the skeleton, Méliès is performing a theatrical illusion, just as he might have performed it onstage: the woman slips into a trapdoor camouflaged by newspaper glued to its surface, and the cloth is held in place by invisible wires that simulate her shape. But the skeleton appears out of nowhere, with no cloth to cover it. This marks the transition from stage to screen: the skeleton part of the trick could only have been done on film, using the stop-substitution technique Méliès had reputedly discovered at the Place de l'Opéra. "The magical appearance," Frazer writes, "is entirely dependent on the ability of the camera to interrupt and reconstruct time."

The reason Méliès was his own lead actor was that he had to keep his position frozen while the camera stopped and the scene was altered. The difficulties were considerable, and Méliès found that he had an uncanny physical ability to meet the demands of stopped time, to fit his own flesh and blood in with deconstructions that would appear to be entirely mechanical. When directing other actors, he would frequently "conduct" their actions to the sound of a metronome, Maelzel's invention, in order to encourage the most automatic, clockwork-like gestures; and his mind was mechanically oriented too: the sudden skeleton, Frazer comments, "comes from the realm of the imaginary inherent in the mechanics of the camera." In conjuring the skeleton, Méliès was responding to the magical possibilities of this new mechanics: he was not just stopping time; he was inventing it.

So the hiatus between the disappearance of the woman and the arrival of the skeleton is important: she does not turn into it; it arrives later, of its own accord. Yet Fischer is not wrong to suppose that the woman's disappearance leads directly to the appearance of the skeleton, since this is clearly what is thought within the film itself. On a first viewing, the trick is so enthralling, and the action so fast (1 minute 15 seconds in total) that you notice little other than the bare facts, or rather fictions. If you play the film again, however, and slow it down at the moment when the skeleton appears, you notice that Méliès, the magician himself, is shocked by its arrival. He looks surprised, shakes his head, and tries to shoo the skeleton away, as if that wasn't supposed to happen. Clearly, even he thinks he has unintentionally killed the woman, instead of simply making her vanish.

The skeleton, a memento mori, becomes a sign of magic's power over him, or (since this is the filmic part of the trick) the cinema's power over him. There are risks, and in this new medium, he has gone from being the sorcerer to being the apprentice. Quickly, he grabs the cloth and covers the skeleton with it. After a swift "abracadabra," he removes the cloth, and, to his great relief, the woman has reappeared on the chair. He offers her his hand, they take a bow, smiling, and leave the stage.

The Vanishing Lady and *Pygmalion and Galatea* became prototypes for a number of Méliès films, in which women appear or disappear: they start out inanimate and come to life; they are conjured out of thin air, or they are made to vanish in a shower of confetti. *An Up-to-Date Conjuror,* made a year after *Pygmalion,* begins with Méliès onstage, and a life-size mannequin on a table beside him. The doll has sculpted hair and is wearing a ballerina costume; she is so stiff that she keeps falling backwards, and Méliès has to hold her steady. All of a sudden, while holding on to the doll with one hand, he brandishes his other in a magical gesture and turns her into a live woman. The woman performs a little dance on the stage, and Méliès subjects her to the Vanishing Lady trick. He makes her reappear on the table, then picks her up and throws her in the air. As she falls she is transformed into a shower of shredded paper—a trick memorable for its display of the magician's power. The flesh-and-blood woman becomes disposable; she is bleached and featureless; her bulk is reduced to two dimensions and torn into tiny pieces.

As in many of his films, Méliès doesn't seem to know when to stop. Though very short, the movies are packed with tricks, and the magician's mania reveals turns that become more and more strange. *An Up-to-Date Conjuror* unfurls at breakneck speed, and is over within a minute. After the shower of confetti Méliès makes himself disappear and come back again. He gets up on the table and jumps to the ground, but mid-jump he turns into the ballerina, who curtseys, blows a kiss to the audience, and gets back on the table. She jumps, and turns halfway through into the magician. When Marey photographed strips of celluloid depicting animals falling to the ground, it was to show how they could painlessly break their fall. Méliès's jumps are like parodies of science, mid-air sex changes, jokes told in a medium that was also used in medicine. At the end of the movie, Méliès scrunches his body up into a ball, and—as if by a supreme effort of concentration— explodes in a puff of smoke.

In other films, women are animated, as in *The Living Playing Cards* (1905) and *The Drawing Lesson* (1903), or appear out of nowhere en masse, as in *The Magic Lantern* (1903). In *Ten Ladies in One Umbrella* (1903), Méliès stands outside a wooden shed labelled "Galatea Theatre," and turns a piece of black cloth into an umbrella. He makes women appear from it one by one, and turns them all into Greek goddesses, dressed in classical costume and arranged as if in a frieze. In *The Clockmaker's Dream* (1904), a horologist falls asleep and imagines all his clocks turning into women. In *The Wonderful Living Fan* (1904), Méliès presents a giant fan to an actor playing Louis XV. When it opens, there is a live woman on each of its seven folds.

In all of these short movies, women are manufactured, and the tools for their making are the magical mechanics of cinema. It is as if, when put together, the films were one long production line, on the scale of that used by Thomas Edison to fabricate his doll. Sometimes, as in *The Microscopic Dancer* (1902) or *The Lilliputian Minuet* (1905), Méliès would even shrink his women to tiny proportions. But the most graphic and most literal of these films, the one that could be said to sum up the fantasies embedded in all the others, is *Extraordinary Illusions* (1902).

Out of a case labelled "Magical Box" come the ingredients of a woman. Méliès pulls out a veil, a metal stand, two mannequin's legs, a torso with arms, and lastly, a woman's head, which the magician kisses as he lifts it from the box. When all of these body parts are thrown together on the stand, they form a life-sized doll dressed like a dancer. Méliès whispers in the doll's ear, kisses her cheek, picks her up, and takes her across to the other side of the stage; but while he is carrying her, she turns into a live woman. Méliès and the dancer then engage in a sort of duel of magic tricks, a mad, antagonistic anthology of the numbers we have already seen: when he tries to kiss her she turns into an enormous and grotesque chef, stirring a pot of soup. The chef so enrages Méliès that he grabs him, finally, and tears him apart, dismembering him as if he were a cloth doll, and sending limbs flying all over the stage.

So what was once a dismembered mannequin turns, when reconstituted, into a monster of the magician's own making, and must be torn apart once again. At every crucial turn—as Méliès is about to kiss the woman, or exert control over her—the object of desire turns into an object of disgust. She slips away from his every intention, out of his physical and mental grasp; even his magical powers are powerless against her. The film acts out, as a surreal, zany, mechanized fantasy, a crucial passage of *Frankenstein*. Disgusted with what he imagines will become of his female progeny, the doctor looks at his second "half-finished creature" on the operating table, and tears it to pieces.

It was a time when the female sex was under a particular form of scrutiny: while Méliès was performing magic tricks at his theatre, another master performer became known for investigating the mysterious behaviour of thousands of women.

Dr. Jean-Martin Charcot arrived at the Salpêtrière, the largest women's hospital in Paris, in 1862. The Salpêtrière numbered over 5,000 patients then, and there was one doctor for every 500 of them. Over half the women were classified under the umbrella diagnosis of "epileptic," there was a cure rate of a mere 9.72 per cent, and, the year Charcot arrived, 254 women died, of "causes linked to insanity."

Many doctors passed through the Salpêtrière during training, but Charcot saw how much work there was to be done there, and was determined to return permanently. He examined each patient individually, separated the true epileptics from the neurotics, and embarked on an investigation of diseases of the nervous system that was to last the rest of his life. Until then, neurology was not a separate discipline from other branches of medicine; Charcot focused on the motor areas of the brain, and made advances in the study of Parkinson's disease and multiple sclerosis. Often referred to as the father of modern neurology, he became known more popularly as "the Napoleon of neuroses."

He was most famous, however, for the discovery, or the reinvention, of an ancient phenomenon that, in the words of his pupil Sigmund Freud, he "lifted out of the chaos of the neuroses": hysteria. As one contemporary put it, "It was as if hysteria were an exclusively French disease, indeed, a Parisian disease."

Charcot expanded his new territory to suit this specific purpose: he added an electrotherapy unit; a special photographic studio, where Albert Londe adapted Marey's camera technique to record patients' hysterical behaviour; and a large outpatient department, from which he would cull demonstration models for his famous lectures.

The charm of his performance on these occasions stemmed, according to Freud, from the fact that they were improvisations—Charcot had never seen the outpatient before him. Spectators were, Freud wrote, "spellbound by the narrator's artistry," and by "the magic of a great personality." Charcot would often illustrate these talks with magic-lantern slides made from Londe's photographs. Thousands flocked to hear him—not just medical students, but also philosophers, writers, and stars of the stage. Henri Bergson came, as did Guy de Maupassant, Jane Avril, the dancer immortalized by Toulouse-Lautrec, and Sarah Bernhardt. The French feminist Catherine Clément later likened this "psychiatric illusionist" to Méliès, and pictured him orating next to his "female monster." "In silent movies," Clément wrote, "and in stories about hysterics we find the same figures of frightened women . . . it is the same imaginary."

Indeed, apart from the performative nature Méliès and Charcot shared, and the different kinds of magic each effected in his field, Charcot also worked with women who were subject to a certain "automatism," and he also, in a sense, brought them to life.

Some hysterics, Charcot observed, would continually behave as if on automatic pilot, a form of behaviour he termed "ambulatory automatism"; others were subject to fits, muscular contractions, and paralyses. This second type of hysteria, *"la grande hystérie,"* involved a stage in which the original trauma would be acted out. The patient would perform what Charcot described as "expressive mimicry," and "whole scenes" would be enacted. This spectacle would last about fifteen minutes, and could be repeated again and again; it was a kind of theatre. Charcot's diagnosis arose, he said, from the observation of the constant repetition of the same mechanical symptoms, a nontechnological form, you might say, of mechanical reproduction. Freud would later write of the uncanny effect of "epileptic fits, and of manifestations of insanity, because these excite in the spectator the impression of automatic, mechanical processes at work behind the ordinary appearance of mental activity."

Often Charcot would place his patients under hypnosis, and then he would notice the way in which the unconscious workings of the hysteric's mind could be seen in her physical movements: "The movements which, on the outside, represent the acts of unconscious cerebration are distinguished by their automatic and mechanical character," he wrote in 1889. "It is then that we truly see before us the human machine in all its simplicity, dreamt of by La Mettrie."

To counter hysterical paralysis or anaesthesia, Charcot applied metal plates to the numb, or "dead," area of the patient's body. In his electrotherapy unit, he adapted this process to include the induction of small electrical currents. Like Frankenstein, or Méliès, Charcot brought to life areas of the body that had previously been inanimate, and mechanized, so to speak, the people in his care.

The symptoms of hysteria, in which patients became automatic, were reflected in the field of art, or of magic, with the invention of cinema. On film, human beings flickered, were made mechanical, and beyond that, in Méliès's trick films, they could do things with their bodies that were impossible in reality. Just as the relation between the mind and the body was being investigated in Charcot's hospital, as the soma seemed most estranged from the psyche, in Méliès's movies heads and torsos would come apart; bodies would be dismembered, paralysed, or reanimated. His *Mysterious Dislocations* (1901), for example, has a pierrot make his limbs detach themselves so that they can reach objects set at a distance; his *New Extravagant Battles* (1900) shows the opposite motion—body parts coming together of their own accord to compose a man.

In 1895, the year the cinema was invented, Charcot's most famous pupil published *Studies on Hysteria*. A few years later, he wrote *The Interpretation of Dreams;* and Méliès, then in his filmmaking prime, strove, as he put it, "to give the appearance of reality to the most chimerical of dreams." If his films are, in many cases, perfect illustrations of what Freud called the Uncanny, nowhere is this more apparent than in those that feature the magician and his double, one of Méliès's signature tricks.

In *The Mysterious Portrait* (1899), Méliès casually walks onstage and sits on a stool next to a large gilt picture frame. Magically, his double appears in the portrait, also sitting on a stool; the double comes to life, and turns to face Méliès. Méliès offers his double a cigarette; the double stretches his arm out past the frame to take it, and puts the cigarette in his mouth. The pair have a friendly (silent) conversation, and then the self in the picture blurs until it fades to nothing.

Sometimes Méliès would give himself multiple personalities (a syndrome studied by another pupil of Charcot's, Pierre Janet). *The One-Man Band* (1900) is a masterpiece, made by a complicated process of multiple photographic exposures that Méliès likened to a Chinese puzzle. The film begins with Méliès standing behind a row of seven chairs. He proceeds to sit down on each chair in turn; every time he gets up, he leaves a ghostly impression of himself behind. Each of his divided selves holds a different musical instrument; each has a slightly different posture or demeanour. They chat to each other and laugh, and when all seven are seated, the man in the middle stands up and conducts them in a silent performance.

In one of his best-known films, *The Man with the Rubber Head* (1902), Méliès took the idea of the double to new heights: in what might be seen as a comic updating of Descartes's mind/body dualism, Méliès plays with his own decapitated head. The scene is set in a chemist's laboratory. The scientist (Méliès) pulls a severed head, a duplicate of his own, from a box and puts it on a table in the middle of the room. He picks up a huge set of bellows and blows up the head, which winks and blinks confusedly throughout, until it is several times its natural size, filling almost half the screen. Méliès then lets the air out of the head through a valve, and it shrinks back to normal size, looking frightened. The scientist laughs and calls his assistant in to see his new trick. The assistant tries it, and blows up the head a little too far, at which point it explodes into a puff of smoke. The scientist kicks his assistant out in fury and, left on his own in the laboratory, sobs in despair.

Perhaps the most magical of the decapitation films is the earliest: *Four Troublesome Heads* (1898). The crazy excesses of the trick are somehow emphasized by the simplicity of the plain black background, and by the fact that it is limited to a single magical gag, without the unstoppable fussiness Méliès introduced in many of his later films. It is the first film in which Méliès used multiple exposures, covering his head with a black bag when he wanted to appear headless, and calculating how far the film had to be rewound before it was exposed again by counting the number of camera cranks (each turn represented eight exposed frames). His head could be filmed on his shoulders, or on a table (in which case the rest of his body was masked in black); but mid-air, in frames that pass so fast they are barely noticeable, his head is a papier mâché dummy, a doll on its way to animation.

In a black room more reminiscent of Marey's scientific experiments than of Méliès's theatrical performances, Méliès stands between two white tables, dressed in a tailcoat. He removes his head from his shoulders, and places it on the table to his left. For a moment he moves about, headless; but then he waves his arms in the air and produces another head on his shoulders. He looks straight at the audience, smiles, and invites applause. Méliès turns to look at his disembodied head—they speak to each other and laugh. He takes the head from his shoulders again and puts it on the table next to the other one. He swishes his arms, inviting applause (or perhaps he is flailing around in blindness—it is not clear whether this is a gesture of triumph or helplessness). He grows a new head, touches it, and laughs, looking directly at the audience. Each of the three Méliès heads wears a different expression, indicating that they are not mere duplicates, but have a life of their own. He places this third head on the other table, quickly grows another one, walks towards the camera, and turns, his tailcoat fanning out, to look at his doubles. He goes to get a stool from the background, and a banjo that has been lying on top of it. Bringing them forward, he sits between the tables and starts to play. His heads accompany him, but after a second or so Méliès covers his ears: his heads are singing out of tune. He gets up and smashes the two on the left with his banjo; they disappear immediately, without a trace. He removes his own head and tosses it offstage like a ball. He grabs the last one left and throws it in the air. The head lands on his shoulders; Méliès laughs, bows, and dances around, almost Chaplin-like. He comes towards the camera, grinning, and walks into the lens until he has become an unrecognisable blur.

The years 1901–1904 were the best of Méliès's career; by 1905, when his theatre celebrated the centenary of Robert-Houdin's birth, things were already going downhill. He had opened a New York branch of his company, Star Films, two years earlier, under the direction of his brother Gaston, but this enterprise proved a disaster. Thomas Edison, who, as we know, had failed to be credited as the inventor of cinema simply because he had not patented his Kinetograph abroad, now wrought his revenge. He claimed that any film made in the U.S. with equipment based on his inventions was an infringement of his patent rights, and after a number of lawsuits he succeeded in gaining an effective monopoly on film production in America. In 1909, the Motion Picture Patents Company (MPPC, also known as the Edison Trust) was established. It acknowledged Edison's patent rights, and granted licences to film-producers and exhibitors, in exchange for royalty payments of half a cent per foot of film. In order to secure these royalties, the seven original licensees, who included Méliès, were required to produce a certain quota of projectable film—1,000 feet—per week. Licences were withdrawn if the requirements were not met, thereby sending the offending companies into bankruptcy.

Méliès churned out thoughtless footage from his Montreuil studio, just so that the company could meet Edison's demands. He tried to make slapstick comedies that he thought would appeal to an American market, but they turned out to be far less popular than those already produced in the States. Gaston, meanwhile, tried his hand at Westerns, and set off on an enterprising journey to the South Seas, where he intended to make documentaries and historical, imperialist dramas; but he met with considerable difficulties. Firstly, he found that the natives of Tahiti were far too civilized to be the subjects of his planned anthropology; then his cameraman took to drink, and his star contracted syphilis. Gaston ended up sending six of his crew members home.

Even more catastrophically, the film Gaston sent back was invariably useless—some was damaged by humidity, some was ruined because he couldn't find enough running water to wash it before he put it in the can, and another batch was destroyed in a shipwreck. Before long, the Edison Trust was suing Star Films for failing to produce its quota. Gaston's son Paul got out of a tight situation by signing over the company, including its stock of films produced by Méliès between 1903 and 1909, to Edison. Gaston's travels cost Star Films $50,000; he never saw his brother again, and died in Corsica in 1915.

Another development, in France this time, had ruinous consequences for Méliès. The International Congress of Film Editors decided that from then on all films would be rented to exhibitors rather than sold outright. This benefited large corporations such as Gaumont and Pathé (which in 1909 was the biggest film-producer in the world), but Méliès didn't have the capital to deal with these complicated arrangements. By 1911 he made films the only way he could—by borrowing money from Charles Pathé with his Montreuil home as security. Méliès's last six films were made under this arrangement, but his fairy-tale compositions proved unpopular and anachronistic in an era that was soon to witness D. W. Griffith's *Birth of a*

Nation. Pathé was unable to recoup the money, and Méliès was soon inextricably in debt to him.

In 1913, Méliès's wife Eugénie died after a long illness. By then, his film production had ceased altogether, and the outbreak of war the following year meant that the Robert-Houdin Theatre, his only remaining source of income, had to be shut down. In dire financial straits, Méliès wrote to John Nevil Maskelyne, asking if he wanted to buy the ten automata Méliès had inherited from Robert-Houdin's widow. Maskelyne turned the offer down.

Many years later, Méliès donated the machines to the Musée des Arts et Métiers, which had been set up in honour of Vaucanson; but when he asked why they had still not been exhibited after four years, Méliès received a stiff reply from an administrator there, explaining that they had been stored in an attic. The attic, the letter said, was subject to extremes of temperature, which had damaged the automata, and eventually a leak had caused the roof to cave in and a beam from the ceiling had smashed them all to pieces. Méliès relayed the news to a conjuring friend of his, mourning "those masterpieces that I jealously maintained in good working order for 36 years." "Who would have thought," he wrote, "and above all in a museum, that they would meet this lamentable end?"

During the war, Méliès turned the buildings at Montreuil into a hospital for the war-wounded, and later into a variety theatre, where he performed shows for the soldiers with his children, Georgette and André. Meanwhile, round the corner from the Robert-Houdin Theatre, Méliès's office at the Passage de l'Opéra was occupied by the military. A number of the films kept there were melted down—the silver was reclaimed and, in an ironic twist on the profession that had made the Méliès family their money, the celluloid was used to make boot heels for the army.

The end of the war provided little relief for Méliès, and 1923, in particular, proved to be a terrible year. The Robert-Houdin Theatre was demolished to make way for the extension of the Boulevard Haussmann; at the same time, Charles Pathé, who had pursued Méliès for over a decade, finally caught up with him. Pathé succeeded in obtaining an order for the compulsory sale of the Montreuil property, in order to repay Méliès's debts. It had been Méliès's family home for sixty-one years, and was the site of the first, now dilapidated, film studio. Before he left, he wilfully eradicated even more of his past. His granddaughter Madeleine, who was born that year, later wrote in her biography of Méliès that, in a fit of frustration, he made a bonfire in the garden at Montreuil and set light to all the films in his possession.

Méliès scraped a living over the couple of years that followed by touring provincial theatres and seaside casinos with his conjuring tricks. He and André worked for a few months restoring the stage machinery in someone else's war-torn theatre, when Méliès could no longer rescue his own; but they were nothing more than hired hands, and on his return to Paris, Méliès saw little future.

He was saved by a leading lady who had not vanished after all. Fanny Manieux, known to theatre and film audiences as Jehanne D'Alcy, had been Méliès's collaborator since he first took over the Robert-Houdin Theatre, and she was also his sometime mistress. In Madeleine Malthête-Méliès's description, there was little love lost between them, and

Manieux was certainly no longer the fairy-like star she had once been; nevertheless, the couple celebrated their marriage at the Hôtel Lutétia in Paris in 1925, and set up shop in a little concession Manieux happened to have—a toy stall at Montparnasse station.

In retrospect, it might seem a fitting end for a man who brought dolls to life, and who made a playful magic out of science. Paul Gilson, a film critic and a regular visitor to Méliès's stall in Montparnasse, described these occasions some years later:

> We would have coffee, one black and one white, at the station bar, and, as I listened to him, our surroundings of fake marble and nickel balls and cigarette butts stubbed out in the sawdust became, to my eyes, a Palace of Mirages. I would not have been surprised to see, at the sound of his voice, the automata of Robert-Houdin sit down on the leather where so many men had left the traces of their thoughts. I liked him, for with him I always recaptured the colours, the charm, the nostalgia of a lost childhood.

Méliès, however, regretted this fate bitterly. He wrote to the film historian Merritt Crawford, in a letter now preserved in the archives at the Museum of Modern Art in New York, that "you can well imagine that the sale of my candies and toys . . . is of mediocre interest to me. I am trying to make a living but what terrible torture for the father of motion pictures to be confined to inaction in a little booth when he is still of active nature and could do many other things, but unfortunately old men aren't wanted anymore." In his memoirs, he described the toy shop as "a prison," writing that it had to be open from seven in the morning to ten o'clock at night. He never had a single Sunday or evening off, he complained, and was forbidden to leave it, even for meals.

Nevertheless, he soon had a steady stream of visitors, amongst them a number of film critics, and others in the industry, who were intent on rehabilitating him. Jean Mauclaire, the founder of the Studio 28 cinema, where Luis Buñuel and Salvador Dalí's *Un Chien Andalou* was first shown, planned to screen what films of Méliès's he could find. Entirely by chance, he found many more than he could have hoped for.

While he was driving in Normandy, Mauclaire's car broke down. The mechanic who repaired it saw the cans of Méliès films that Mauclaire had on the back seat, and commented that some cans just like that had been found in a dairy at a nearby chateau. The chateau had once belonged to Dufayel, a department store owner who had opened a cinema in one of his shops, and had been one of Méliès's most regular clients. Mauclaire made his way there as soon as his car was fixed, and found no fewer than eight Méliès films. He had them restored in Paris, and coloured by hand according to the precise instructions of Mme. Thuillier, who had once coloured Méliès's films herself.

On 16 December 1929, a gala was held in Méliès's honour at the prestigious Salle Pleyel, and his work was introduced to a new generation. Afterwards, the surrealists—André Breton, Louis Aragon, Paul Eluard—signed a sketch of him posted up in a Montparnasse brasserie; in America D. W. Griffith and Walt Disney pronounced themselves fans. The films Mauclaire had gathered together were shown, and there was a short, specially made for the occasion by Paul Gilson. The audience had just seen Méliès in his films, when he was in his forties. Now they saw an older version, a man of sixty-eight. Madeleine Malthête-Méliès, who was at the gala with her parents, described the short film and the performance that ensued:

> Lost in the streets of Paris, he looks everywhere for the Salle Pleyel. Bits and pieces of film hang out of his pockets. He is anxious, nervous, and the strips of film fall down so they make him stumble. He looks around, turns right, turns left. Now he is nothing but a tangle of film stock in which he almost disappears. Almost overwhelmed by the celluloid, he sees a huge announcement of the Gala, with a big caricature portrait of himself. Here he is. The lights are turned on in the theatre. The screen is raised, showing, in the middle of the wall, a frame with the poster we have just seen. Suddenly the paper breaks in the middle, and Méliès shows up, alive and well, in evening clothes, under the light of all the reflectors. He is cheered by an endless ovation.

Gilson's tribute film is a perfect emblem of Méliès's work: the magician is almost overwhelmed by his own mechanical reproductions, but just in time, he finds his double in a poster. The poster, in turn, finds its double on the stage, and the two-dimensional pen-and-ink Méliès is torn through to reveal the man of flesh and blood, come to life.

All of Méliès's family came to the Pleyel Gala, but one of them, Georgette, had already contracted the illness that was to kill her within months. In 1930, Méliès wrote to Merritt Crawford that "the death of my unfortunate daughter, who was but 42 years of age, caused me much sorrow. She was a pleasant woman and a lyrical artist of high order. At all times since the death of her mother in 1913 she was a very able and devoted assistant; this applies to her work at the Robert-Houdin Theatre, and our theatre of Artistic Varieties at Montreuil, and to my cinematographic work. She was the first woman projectionist and the first 'camera woman'; also one of the first motion picture actresses."

Georgette's widower, a baritone singer in comic operas, was always on the road, so Méliès and Fanny took charge of Madeleine, who was then seven and a half. She moved in with them, and slept in a separate small apartment next door, along with all the toys that Fanny stored there. Méliès would wake her up in the mornings by knocking on the wall between their bedrooms and, now in his seventieth year, would take the child to school. She would have lunch with her grandparents at Montparnasse, where Méliès sketched constantly, and Fanny cooked on a camping stove at the back of the toy stall. Merritt Crawford's sister visited them there, in "what [Méliès] calls his cage," and wrote to her brother:

> I wish I could give you a picture of that inane, cheap little booth, with its flimsy toys and gay wrapped candies, artificially built into a kind of proscenium arch, behind which one sees a family picture—Madame

with her little cooking outfit, presiding housewifely over the cuisine, and at the same time handling the customers, and the little granddaughter who lives with them, pale and quiet on her stool, and the fine sensitive head of Méliès bent over his work.

Despite the many visitors and fans Méliès received at this toy theatre, he was desperate to end his days somewhere better than the freezing stall; and yet, even after he was awarded the Légion d'Honneur, after he received a state pension and was given an apartment on the outskirts of Paris, in spirit Méliès never really left the world of toys.

Years before Méliès took up his work in the Montparnasse stall, the French poet and critic Charles Baudelaire had written of how much he enjoyed walking past Paris toy shops. There, he claimed, he found "the whole of life" in miniature—the sparkling eyes, the rosy cheeks; the drunkards, the bankers, the clowns; the gardens, the theatres, the armies. For children, he suggested, there was something even more lifelike about these theatrical scenes: "All children talk to their toys," Baudelaire wrote, "the toys become actors in the great drama of life, scaled down inside the *camera obscura* of their young minds." Méliès had spent his life projecting the camera obscura of his mind on to a bigger screen, shrinking people to miniature sizes, filming fairy tales and magic tricks, bringing dolls to life. He died of cancer in 1938; those who knew him often said that the old man had never ceased to be a little boy.

CHAPTER FIVE

The Doll Family

All moveables of wonder, from all parts,
Are here—Albinos, painted Indians, Dwarfs,
The Horse of knowledge, and the learned Pig,
The Stone-eater, the man that swallows fire,
Giants, Ventriloquists, the Invisible Girl,
The Bust that speaks and moves its goggling eyes,
The Wax-work, Clock-work, all the marvelous craft
Of modern Merlins, Wild Beasts, Puppet-shows,
All out-o'-the-way, far-fetched, perverted things,
All freaks of nature, all Promethean thoughts
Of man, his dullness, madness, and their feats
All jumbled up together, to compose
A Parliament of Monsters.

—William Wordsworth, *The Prelude*

Nothing is more easy than to err in our notions of magnitude.

—Edgar Allan Poe, "Maelzel's Chess Player"

In 1922, when they were hired as a song and dance trio at Coney Island, Frieda, Kurt, and Hilda Schneider decided to call themselves "The Dancing Dolls." They were midgets, or hypopituitary dwarfs—perfectly in proportion. People said they looked like dolls, and so that is what they became. They were literally "living dolls": humans in their ordinary lives, but dolls when it came to making a living. Frieda was twenty-three, Kurt was twenty, Hilda was fifteen; the two eldest were about three feet tall, Hilda was just under four.

It must have seemed a simple enough decision to them— little people were often referred to as "dolls" at the turn of the last century. At the circus, "dolls of the sawdust" was a familiar nickname for sideshow midgets. The performance of any sideshow midget depended on a certain perplexed curiosity among the crowd: were they adults or children? Had they been miniaturized by some unfathomable scientific process? Age and size were called into question: midgets befuddled, fundamentally, their viewers' sense of time and scale.

Yet in calling themselves dolls, and inviting their audiences to see them as a cross between the animate and the inanimate, the Schneiders were voicing something far more strange to their audiences. Were they still objects, brought to life? Were they mechanical? Were they born or manufactured? Their performance was not just a song and dance act, in other words: it was a balancing act, treading the fine line along each of these borders of perception. When the Schneiders called themselves dolls, they seemed to be announcing that the Uncanny was their human—or at least their professional— condition.

In his essay on the Uncanny, Freud wrote that children are

more likely to want their dolls to come to life than they are to fear it. Charles Baudelaire thought that "most children want above all to *see the soul* of their toys," continuing:

> I cannot blame this infantile mania, it is the first metaphysical stirring. When this desire has become fixed in the child's brain, it fills his fingers and nails with an extraordinary agility and strength. The child twists and turns his toy, he scratches it, shakes it, bangs it against the wall, throws it on the ground. From time to time, he forces it to resume its mechanical motions, sometimes backwards. Its marvellous life comes to a stop. The child . . . finally prises it open . . . But *where is its soul?*

Vaucanson, Edison, Méliès, and the people who saw the Chess Player—who wanted to believe in it, and were blind to the trick—all shared these "metaphysical stirrings." If they couldn't find life in a toy, they wanted to manufacture it, or see it though they knew it wasn't there. Even Descartes, that rational theorist of the soul, imagined his doll as his daughter; the sailors who accompanied him believed she was the bearer of bad luck. Whether out of sentimentality, mania, or superstition, a doll never exists merely as a physical fact. A doll is always also a figment of the imagination. Its life is in the eye of the beholder. And so, in implying that they had been made instead of born, the Schneiders' performances told a certain truth. Though they weren't invented by a mechanic or a philosopher, the Dolls were someone's creation: they were projections, constructions, whatever their audiences made them.

What their audiences thought they were, sometimes at least, was mechanical dolls. In 1934, the *New Yorker* ran a series of articles about the circus written by Alva Johnston. By that time the Schneiders had been joined by another sister who was also a midget. They had become part of the Ringling Brothers troupe and were known as "The Doll Family." Johnston reported that audiences often felt they were being cheated, since they couldn't believe the midgets were human, and insisted they were being swindled with a set of mechanical dolls. Whether this was because the Dolls looked so pretty and sweet, or because people just didn't think humans existed in that size, it's hard to tell. Certainly, they don't appear to have simulated any mechanical movements as part of their act, to have cultivated this reaction especially. Yet it seems clear from the audience's indignant response that what caused amazement in public spectacles had shifted over time.

Charles Babbage, the nineteenth-century mathematician whose inventions are now regarded as the forerunners of the computer, mourned precisely this change, in his autobiography. Babbage had exhibited part of his "Difference Engine" at the Adelaide Gallery in 1832. A decade later he tried to raise the money he needed to construct his next calculating machine by making a small automaton. "I imagined that the machine might consist of two children playing against each other," he wrote—the game he envisaged was noughts and crosses— "the child who won the game might clap his hands . . . after which, the child who was beaten might cry."

Although this was the sort of object that would have flourished in Kempelen's or Maelzel's time, Babbage found there was no longer a market for it. The public didn't want to see mechanical children; it wanted child-simulacra of another kind. At the Adelaide Gallery itself, "the most profitable exhibition which had occurred for many years," Babbage wrote, "was that of the little dwarf, General Tom Thumb."

Phineas Taylor Barnum, the impresario behind Tom Thumb, led the way for human oddities to replace mechanical curiosities in the public imagination. Even before then, that fascination had been in place, and it was often confused with an appetite for mechanics. Twenty years earlier, William Jerdan, the editor of the *Literary Gazette,* had paid a visit to Caroline Crachami, the "Sicilian Fairy" on exhibition in Bond Street. Just as the fictional Lemuel Gulliver reported that the King of Brobdingnag "conceived I might be a piece of clock-work . . . contrived by some ingenious artist," Jerdan marvelled that "the machinery of life could [be] carried on in so minute and diminutive a form." "It is impossible to describe the miracle of her appearance," he went on, "or its effect upon the mind . . . A tolerable sized doll, acting and speaking, would not astonish us so much."

The Uncanny, in other words, had reversed its aspect. Instead of wondering if automata were human, people now asked themselves how such purported humans could contain the requisite "machinery." When Kempelen toured with his Automaton Chess Player, audiences believed the machine could only work if there was a dwarf hidden inside it. By the time the Doll Family were in their prime, fully visible dwarfs were considered so amazing that people wanted to believe they were machines.

The year 1914, when Frieda and Kurt Schneider left Germany, was the beginning of the end for the actual doll industry in Europe. The First World War brought with it a shortage of materials and staff, and several large factories shut down. The famous nineteenth-century signatures of Jumeau and Steiner, the porcelain dolls given to well-off children, the Victorian idyll of the nursery—all that was replaced with more mass-produced fare. Soon dolls joined human émigrés, and America became the centre of production.

Until then, however, Germany and France had led the world in doll manufacture. It was Germany's traditional pre-eminence in this field that led a late-nineteenth-century journalist, Edouard Fournier, to confuse some midgets who visited Paris with dolls. "They arrive from Germany," Fournier reported, "just like toys . . . They are as tall as a hardback volume of *Gulliver's Travels* . . . They are well-formed, even their heads are well-proportioned; all in all they are true men reduced to statuettes . . . They come, they go, they gesticulate, they speak . . . You expect a toy, an automaton, and this way you have it whole; with another accent the object would seem improbable."

He might just as well have been describing the Doll Family, since this expectation was what they played on with their name. By the beginning of the twentieth century, dolls were mechanically very sophisticated. Edison's phonograph had been appropriated by at least two European manufacturers, and others made their dolls talk by means of the bellows patented by Maelzel in the 1820s. When Kurt Schneider was born, dolls could walk, talk, laugh, cry, suck on a bottle, simulate sleep and breath, and even—Vaucanson would have been amused to know—"digest" their food. So when the Schneiders said they were dolls, they did mean, implicitly, that they were mechanical.

What had disappeared, as Charles Babbage noted, was the "philosophical toy." While the eighteenth century had sought to enlighten the masses with science and ideas, the technological inventions of the nineteenth century became a rowdier sort of people's spectacle. There was something carnivalesque about the beginnings of cinema, for example, that never arose even around such potentially vulgar eighteenth-century exhibitions as a defecating duck. The marvellous had become anarchic. By the twentieth century—the century of mass production, of working classes and leisure after hours—it seemed the worlds of science and amusement had taken resolutely forking paths. There was no Diderot of the pleasure park, and no Méliès of the calculating machine. Science was a matter for private study, not public places.

Our story—the story of mechanical simulations and distractions, of mad beliefs and curious spectacles—continues here, however, in the colourful, quizzical world of the circus. Its audiences are the heirs to those of Vaucanson and Kempelen; to the people who believed Edison was capable of anything, to those who ran from cinemas in fear. "Look around you," read the preface to a private scrapbook of playbills and advertisements from the eighteenth and nineteenth centuries, "and contemplate the various Shows and Diversions of the People, and then say whether their temper of mind at various periods of our History, may not be collected from them?" The "temper of mind" of the public, over the years, has been in part a quest for that which defies the mind, which is beyond belief; and here, instead of a doll that seems to be human, we have people who seem to be dolls. Their lives might be thought of as a metaphor—the story of countless automata inverted, a biography of the Uncanny.

Frieda Schneider was born in Dresden in 1899, and her brother Kurt was born three years later. Their father was a cabinet-maker, a man nearly six feet tall. Kurt grew normally until the age of seven; he realized he was much smaller than his friends when, as he told a reporter from *Collier's* magazine many years later, he could not reach the fruit they were stealing from the neighbours' trees. Since his sister was of a similar size, their parents took them to see a series of doctors. They saw seven in total, and the medical advice they were given ranged from eating plenty of vegetables to a suggestion that the children might be stretched on a rack. Eventually, the Schneiders accepted the fact that they were midgets.

Kurt considered becoming an apprentice to a watchmaker (like Robert-Houdin or John Nevil Maskelyne), so that his tiny fingers could be put to an appropriate use; but his father recommended that he and Frieda make a living as an exhibition. So, instead of making mechanical things, he made an object of himself.

The history of midgets and dwarfs, at carnivals and at court, is that of humans treated as objects. Just as Gulliver was appropriated by the Queen of Brobdingnag as a pet or a toy, dwarfs outside fiction were seen not only as inanimate things brought magically to life, but also as objects to be possessed. The eighteenth-century dwarf Joseph Boruwlaski played on this fact when he signed his love letters "your toy"; but he suffered an ironic indignity when that lover jilted him and left him on her mantelpiece, where he flailed helplessly amongst her porcelain figurines. Edward Wood, author of a history of giants and dwarfs, quotes a handbill from 1712, advertising "a little Black Man being about 3 foot high, and 32 years of age," and his wife, "commonly called the Fairy Queen," who rode a turkey as their horse. What was exceptional about these exhibitions, the handbill pointed out, was that they were "all alive"—as if they had been animated by magic, and this Promethean aspect were the point of the show.

In 1835, Charles Dickens saw a dwarf at Greenwich Fair, which he described as "an unfortunate little object." Caroline Crachami, who had been compared to a doll in her short lifetime, was carried about like one in death: the man who had exhibited her tried to sell her body. He offered it to various hospitals, until it was accepted by the Royal College of Surgeons, who prepared and articulated her skeleton—that once-living object remains in a glass cabinet, in a public museum, to this day.

The Schneider children toured Germany for a while, performing a song and dance routine called "Hansel and Gretel" in nightclubs, and in 1913 they met an American couple named Mr. and Mrs. Earles. Accounts of this meeting differ: some say that Bert Earles spotted the Schneider children at the international freak market that used to take place every year in Hamburg; a more reliable version suggests that his wife was herself a performer—a German giantess over seven feet tall—who met them in a cabaret where they were all performing. Her husband Bert was a businessman, and had nothing to do with the circus. Mrs. Earles became their mentor and took them back to America with her when she finished her tour; Kurt and Frieda went to live with the Earleses in Pasadena in 1914. They changed their names to Harry and Grace Earles, and found work as part of Bill Cody's "Buffalo Bill's Wild West Show."

Meanwhile, the Schneider parents continued to have children. Margarete and Rosa were both of normal size; a fifth child, Hilda, was a midget; Dorothea, Max, and Elizabeth were all tall, and the Schneiders' last child, Elly, was also a midget. There were nine children in total, seven of them girls. After the First World War, Mrs. Earles returned to Germany and brought Hilda back with her to America. Hilda took the name Daisy, and formed a dance trio with her brother and sister. Between 1922 and 1925 they decided to try their luck on the East Coast, where they were soon hired as performers at Luna Park in Coney Island.

Luna Park was built by a man after Georges Méliès's heart. Frederic Thompson was an architectural draftsman who had dropped out of the Ecole des Beaux Arts in Paris and made his name with a funfair ride called "A Trip to the Moon" in the same year as Méliès's film of that title came out. Once he and his business partner had taken a twenty-five-year lease on a 22-acre seaside property in Coney Island, they added other Jules Verne– and Méliès–inspired rides, such as "Twenty Thousand Leagues Under the Sea," and Thompson set about designing the buildings.

As Thompson later explained in "Amusing the Million," an article he wrote for *Everybody's Magazine,* the secret to a great theme park was "the carnival spirit"—in other words, "speed, light, gaiety, colour, excitement." "When people go to a park or an exposition," Thompson advised, "and admire the buildings . . . without having laughed about half the time until their sides ached, you can be absolutely sure that the enterprise will fail." Thompson constructed Luna Park out of mad minarets and spires, until he had a total of 1,221 red-and-white-swirling towers, lit up by over a million electric light bulbs. At a time when electric lighting was not all that common, Luna Park consumed more energy than the average American city. "Straight lines," Thompson declaimed, "are necessarily severe and dead"; he said architecture was his "freak show," the outside attraction that drew the public in.

Though in the daytime Grace, Harry, and Daisy Earles wore children's clothes (Harry's shoes were 8½, baby size), for their performances as "The Dancing Dolls" they dressed in elegant cabaret costumes: Harry wore a top hat and tails made to measure by a tailor; the girls wore long satin dresses sewn by Daisy, and high-heeled shoes made specially for them in New York, or shipped from Chicago. (Many years later, these costumes were given away as dressing-up clothes for children.) They looked like miniature Tallulah Bankheads, with Harry as their matinee idol.

Their act was just one part of the zany atmosphere Frederic Thompson hoped to create. Over time, he filled his theme park with elephant rides, shipwrecks, a miniature subway, and a reconstruction of Niagara Falls; there were performing monkeys, helter-skelters, Babylonian hanging gardens, trapeze artists, acrobats, lion-tamers, and clowns. "In every part of the grounds something extraordinary was going on all the time," he wrote,

Instead of advertising an organ concert in the Music Hall we yelled ourselves hoarse about high diving, greased poles, parades, and every other crazy thing we could think of . . . To the stadium, which had never held a quarter of its capacity, I drew 23,000 people to see a race contested by an ostrich, a camel, an elephant, a man on a bicycle, another on a horse, an automobile, and a zebra. I had a man sliding by his teeth from the top of the sky-scraping electric tower to the esplanade below. True, he had never before traveled more than thirty feet in that fashion, but . . . the illusion was great, and the stunt made a sensation.

Elly Schneider was born the year her eldest sister and brother left for the States. She had never met them when, at the age of eleven, she too became part of their act. Mrs. Earles came to find the last of the Schneider midgets in Germany and, without warning the others of her arrival, introduced Elly to her siblings. Around that time, they joined the Ringling Brothers Circus, giving their first performance at Madison Square Gardens. They became known as the Doll Family, and Elly took the name Tiny Doll. Since she spoke no English at first, playing a silent doll was exactly what she was required to do during the shows, though later she learnt to ride a pony, and to sing "I'm Looking Over a Four-Leaf Clover," and other songs they regularly performed.

The Ringling touring season was from April to November. According to a circus programme from 1933 (the Ringlings' golden jubilee), the show travelled almost 20,000 miles in a single season, and gave over 400 performances. The programme gives details of other astronomical figures involved. Forty-eight hundred meals were cooked a day, for example, and it was "not unusual for 10,000 pancakes to be baked and eaten at a single breakfast." The daily shopping list for the 1,500 employees, 735 horses, six herds of elephants, and 1,000 other wild animals included: "300 lbs. of butter, 300 gallons of milk, 200 lbs. of coffee, 35 bags of table salt, 2,500 lbs. of fresh meat, 2,000 loaves of bread, 250 dozen eggs, 1,500 lbs. of vegetables, two barrels of sugar, 50 lbs. of lard, 100 dozen oranges, 50 tons of hay, 20 tons of straw, 350 bushels of oats and four cords of wood." There was a post office, and the circus had its own doctors, lawyers, dentists, and "force of detectives."

The Ringling train was made up of more than a hundred double-length railroad cars. Walker Evans, who photographed the Ringling railroad cars and circus wagons at their winter quarters in 1941, showed them to be decorated in the most extraordinarily baroque way. Some carried carvings of winged mermaids whose tails ended in a flurry of exaggerated swirls; there were knights on horses reminiscent of those on Florentine tarot cards, gilded jesters, and half-dressed goddesses wearing helmets.

The Doll Family had a stateroom on the Ringling train that ran the whole width of the sideshow personnel's Pullman. The room contained a small toilet and washbasin, and three lower berths for the three girls. Harry thought his upper berth a waste of space, and had a smaller shelf built instead. The shelf measured 45 by 22 inches, and was attached by a hinge, so that it could be folded back against the wall. He reached it via rope ladder.

There was a 5-by-8-foot dance floor in the sideshow performers' car, where they danced to phonograph records; but Harry was mostly to be found playing blackjack in the evenings, and smoking a cigar. One of his best friends was Jack Earle, a giant known as "Sky High," who, according to one description, stood "almost as much more than six feet as Harry is less." They were often photographed together, Harry perched on Jack's shoulder, or standing on the ground, reaching just above his knee. In the afternoons, the two of them would sneak off for a walk around the busiest roads in town and cause a spectacle by loudly calling each other "Daddy" and "Junior."

According to one of Alva Johnston's *New Yorker* articles, the front seat of the bus that took the circus freaks from the railroad to the big top was reserved for Jack Earle and the midgets, thus giving them "a social superiority." The Dolls never considered themselves to be exactly like the others: in the circus, they were treated like sideshow royalty.

There were things to watch out for all the same. When he was interviewed by *Collier's* magazine in 1942, Harry listed the occupational hazards, in descending order of danger, as: "storms, drunks, women, smart alecs and dopes." He had learnt his lesson about the first of these when a strong gale blew the dining tent down. Harry was knocked on the head by a pole, and was stuck, unconscious, under the fallen tent for over an hour before anyone found him. Even a brisk breeze, he told the reporter, could sweep a midget right off his feet. Harry had often been hassled by drunks much bigger and stronger than him; he hated the wise guys who thought calling out "Hey, Shorty!" or "Hello, Shrimp!" was funny, and as for women, he said: "Do *you* want to get snatched up and cuddled into 200 pounds of perspiring fat? Well, I don't either!"

Another risk to any circus midget's livelihood was growth. Unlike normal-sized people, whose bone-ends close in their early twenties, midgets can suddenly shoot up at any age. One midget, Eddie Wilmot of Minneapolis, is recorded as having reached six feet by the time he was twenty-eight. The Doll Family had not grown much in recent years, and they remained about the same height all their lives. Harry was three feet tall, as was Grace. Tiny was two inches shorter than that when she arrived in America, and later grew another five inches. Daisy, the tallest of the Dolls, was four feet, tall enough to drive the family car, using extension pedals. The maximum height for a midget was given by one circus magazine as four feet eleven and a half inches. Daisy was, according to the *New Yorker,* "on the borderline between a tall midget and a short woman"—given another few inches, she could "outgrow the freak classification entirely." The Ringling Brothers promised her that if she grew another inch or two they would make her a snake-charmer instead.

But Daisy never grew that much, and the Doll Family remained a foursome with the Ringling Brothers for the next thirty years; they became, in the words of the circus newsletter, *The Bandwagon,* "one of the most talented and most famous midget groups in all present day history."

During the French Revolution, the last court midget in France became a royal spy. Richebourg disguised himself as a baby and, carried in and out of Paris in the arms of a nurse, smuggled secret messages under his infant's cap. A couple of centuries later, the very same disguise earned Harry Doll a career in Hollywood.

Harry Earles (as he was credited) was given a number of roles in which he played infants. There was a spate of them in the late twenties: *That's My Baby* (1926), *Baby Clothes* (1926), and two Hal Roach comedies starring Oliver Hardy—*Baby Brother* (1927) and *Sailors Beware* (1927). The idea was always that the baby, in the manner of Richebourg the spy, would turn out to be more than a baby, or to have an uncanny ability to suddenly behave like an adult. Once again, a Doll family member managed to play on these borderline effects as a kind of running joke.

Midgets were sometimes recruited where children seemed unmanageable or, ironically, not childlike enough. Few parents of the twenties knew, for example, that the sweet little infant whose face adorned every jar of baby food and bottle of talcum powder was Franz Ebert, an Austrian midget who had toured America with the impresario Leo Singer. But the joke with Harry and the slapstick stars was different: it was an intentional double-take. And whatever its comedy value, his employers had taken the idea from a horror film.

In 1925, the year before these comedies began to appear, Harry had starred in what was to become a classic of the crime genre, *The Unholy Three*. It was directed by Tod Browning, a man known as "the Edgar Allan Poe of the cinema." Browning eventually became best known for *Dracula*, with Bela Lugosi, for the films he made with Lon Chaney, "the man of a thousand faces," and for his controversial *Freaks*.

Based on a bestseller of the same name written by Tod

The Doll Family

Robbins and published in 1917, *The Unholy Three* follows three circus performers who, dissatisfied with their lives and embittered towards their audiences, embark on a life of crime. Harry plays Tweedledee, a performing midget; Lon Chaney is Echo the ventriloquist (slightly limited by the fact that the film is silent); the third character is Hercules, a strong man played by Victor McLaglen, a former boxer. The trio set up a pet shop, where Echo's multiple voices make grizzly-looking parrots seem magical or spooky—they have conversations with each other about the customers, quote lines of poetry, or recite wise aphorisms (in the book, one of them is described as "a feathered Schopenhauer"). Echo is transformed into a harmless little old lady, and Tweedledee is the adorable grandchild she pushes along in a pram. (Hercules, the heavy, remains hidden until his services are required.) The routine is to get customers to pay for a pet, which the old lady will deliver to their home later. Once there, Echo and his infant case the joint, and return to rob the customers at night, with Hercules as back-up.

The studio, MGM, didn't think it sounded all that promising. They imagined its effect would be less frightening and more the kind of thing Laurel and Hardy used Harry Earles for later. "You'll never get audiences to take a delinquent disguised as an old lady and a midget passing for a baby seriously," they told Browning. "That's the plot of a comedy. Mack Sennett could use it and get people to howl with laughter, but in a mystery it would be impossible."

They weren't entirely wrong, since *The Unholy Three*'s success as a subtle horror movie relies on the ambiguity of the disguises. What makes them creepy is precisely the fact that they border on the funny: you're never sure whether to laugh

229

or wince in fear. This is particularly true of Harry's performance. Chaney is simply a man in drag, whereas Harry fuses unlikely stock characters in an eerier way. He looks the most authentic gangster, a miniature James Cagney with a trilby and a rumpled face. He is more aggressive than his larger partners in crime, and yet his size makes him a mastermind who can never be followed or traced, who slips through the transoms above doors. His face has the look of a fierce adult, while his form, clothed in a long white nightdress, is convincing as that of a baby. He reaches out of his pram at one point and tries to grab some jewels: his eyes sparkle with ferocious greed; he clenches and unclenches his fist in front of a dangling ruby necklace, as its owner bends over the pram to admire him. His appearance on screen turns the theory of the Uncanny into unmanageable fact.

Once he has managed to steal the necklace, Tweedledee hides it in a toy elephant. When the police come to the pet shop to investigate, he plays with the elephant, wearing his nightgown and a toy fireman's hat, and pretends to be a possessive, tantrum-prone child when they ask to look at it. The elephant is an automaton, whose trunk moves up and down as Harry plays with it.

As soon as the policemen have left, Harry gets up, throws his fireman's hat on the floor, bunches up his nightdress to reveal suit trousers underneath, and reaches into his pocket for a packet of cigarettes. He shoves one in his mouth and leans moodily on his elbow against a dresser. At other times, he is seen smoking a cigar, still in his hitched-up nightdress, trouser-clad legs up on a table. While sitting in a baby's high chair, he lifts up his dress and bends down to tie the shoelaces of his black brogues.

On each of these occasions the viewer does a double-take. On the most straightforward level, Harry's double-edged performance is, as the makers of his later comedies knew, a good sight gag; but, more unsettlingly, where the duality should be between adult and child, it also seems to be between child and doll—the quality of imitation, rather than what is imitated, is what comes across most strongly. Flickering across the screen with jagged movements, Harry looks like a copy of a baby. Something tells us, when we look at him, that he is not a real child; but that might be because he is an automaton, rather than because he is a man. As he plays with his toys, one of them a mechanical elephant shrunk down to a tiny size, it is hard to tell from the film whether he might not also be one of them, secretly mechanized. At the end of the movie he taunts a young girl by playing with the ventriloquist's dummy: the dummy is bigger than he is, and he sits behind it, making it move and laughing hysterically. The scene seems horribly impossible—a Doll manipulating a doll.

In the book, the reasons Tweedledee turns to crime are made melodramatically clear. Robbins writes:

> All that he asked was to be taken seriously; and yet no one had granted him this simple wish . . . No, he had been a doll; a plaything for all these vulgar children of the world—children who paid to see him move his head, open his mouth and speak—children quite careless of the inner workings of their doll . . . And, as he

had grown older, the inner workings of this doll had changed; strange transformations had taken place; the springs of good had corroded with rust; and soon the green mould of evil covered everything.

It is for the children in the audience that Tweedledee reserves his strongest spite, since they are too much like him, or rather, he is too much like them. Robbins goes on:

> Tweedledee sat gloomily staring at the gathered crowd, at the round-eyed children—yes, at the children, for these he hated most. They were caricatures of himself. These little brainless beasts had bodies like his own. And because of this he was treated like them . . . Yes, he hated them most. Their piping voices, their pointing fingers, their curious eyes—all filled him with a nauseating hatred hard to bear. At the sight of them, he felt tempted to spring forward, to dig his fingernails into their soft flesh, to hurl them to the ground, to stamp them into unrecognisable bloody heaps.

These violent thoughts were translated into a scene in the movie in which Harry was required to strangle a little girl exactly his height. The girl finds Hercules and Tweedledee robbing her home, and exclaims to Hercules, "Oh, Santa Claus! You've brought me a brother!" But Tweedledee is not amused by her assumption that he is a child. He approaches the girl, wearing his trilby and dark overcoat, with a frown on his face and a pair of stockings in his hand. The girl, an innocent-looking thing with ringlets in her hair and a lace-collared jumpsuit, turns and smiles at him. He reaches out to her, clasps his tiny hands around her neck, and she falls to the ground, choking. The scene can only now be seen in stills; it was thought too shocking, and was cut.

Uncanny pranks were not limited to the screen. On set, Harry joked around with Lon Chaney—"one of the best friends I ever had," Harry later said. Once, Chaney walked into the studio's wardrobe department, still dressed as an old lady, and asked if he could borrow a clean dress for the baby in his arms. "Of course," replied the wardrobe mistress, "why, I'll change the baby for you myself." Suddenly the baby shouted, "Like hell you will, madam! Not *this* baby!" and, twisting out of Chaney's arms, cried: "You dirty, double-crossing bum! That's the last time I'll ever fall for one of your gags!"

While they were shooting, Victor McLaglen used to spike Harry's baby bottle with Scotch. "That was okay by me," Harry later told *Collier's* magazine, "until he and Lon decided they ought to cut in on it. I had to threaten to choke the bum before he'd give it back."

The Unholy Three was voted one of the ten best films of 1925 by the *New York Times*. It was considered such a success, in fact, that when sound came it was remade as a talkie, with exactly the same cast. Lon Chaney, no longer limited to throwing his voice inaudibly, was transformed in the publicity material from "the man of a thousand faces" to "the man of a thousand voices." He was so convincing the studio made him sign an affadavit stating that all five voices were really his own. Chaney was the child of deaf-mute parents; he did not speak for the first several years of his life, and much later, with the advent of talking pictures, he was reluctant to convert to sound, saying it would "destroy the mystery" of silence. There were almost too many ironies in the fact that he played a ventriloquist in what was to be his first talkie and last film: he died of throat cancer the year audiences heard him speak.

Harry was keen to work with Browning again. A few years later he found something else written by Tod Robbins—a short story called "Spurs" published in *Munsey's Magazine*—and passed it on to Browning. The story was about a circus midget called Jacques Courbé who falls in love with a tall bare-back rider. Since Courbé is rich, the woman marries him for his money, but as soon as the wedding is over she makes clear that she is really in love with her riding partner. The tall people think the joke is on Courbé. Dressed up as a cavalier, riding a dog he calls his charger, Courbé's role in the circus is to imper-sonate the bareback riders' daring deeds in a piece of mimicry so ridiculous that banana skins and orange peels are thrown at him from the audience. But Courbé is determined that his mar-riage should not be subject to the same kind of laughs. On hearing of her true allegiance, he makes his new wife his slave. He forces her to carry him everywhere and drives her into the ground until even her beloved rider has forgotten her.

Robbins's story covered some of the same territory as *The Unholy Three:* those who are mocked will have their revenge. Browning turned it into *Freaks*. The film came out in 1932; in some countries it was banned for over thirty years.

Browning was not unfamiliar with the world he set out to portray. As a teenager in Kentucky, he had run away from home and joined a travelling circus. He was, at various times, a contortionist, an escape artist, a jockey, a clown with the Ringling Brothers, and a sideshow talker, rounding up audiences at carnivals for such curiosities as "the wild man of Borneo." At some point in his travels he was hired as an assistant to Leon Herrmann, a French magician who was the nephew of "The Great Herrmann," a regular performer at the Egyptian Hall. On a floating fairground that sailed the Mississippi in the early twentieth century, Browning played a "Living Hypnotic Corpse." He would act as though he had been put into a trance by a hypnotist, fall into a coffin, and be buried alive, sometimes for up to forty-eight hours. He told a journalist in 1921 that the act had given him "time to think."

Browning wanted to make his film the story of more than just one midget, though Harry was to be the hero. In order to produce a show of collective strength by a group of people thought to be helpless, Browning recruited a huge cast of real circus freaks. He hired Prince Randian, the Hindu Living Torso, who could light and smoke a cigarette using only his lips and chin; Daisy and Violet Hilton, the pretty Siamese twins; Joseph-Josephine, the half-man-half-woman; Johnny Eck, the boy with half a torso; Kookoo the Bird Girl, and numerous others. Daisy played Harry's jilted fiancée; Olga Baclanova, a Russian stage actress who had come to Hollywood after many years at Stanislavski's Moscow Arts Theatre, played his full-grown wife; the British actor Henry Victor played Hercules, the man she really loves.

The first part of the film followed Robbins's story almost exactly: Harry plays Hans, who is dressed in black tie—as Harry was in his own acts—rather than in Courbé's baroque attire. He falls in love with Cleopatra, a trapeze artist (in the film, the bareback rider is Frieda, a beautiful midget played with soft emotion by Daisy Earles). The wedding dinner, at which the only non-freaks are Cleopatra and Hercules, is a raucous affair. As a rite of passage into their world, one of the dwarfs jumps on to the table and passes a goblet around, while they all chant at Cleopatra, "One of us! We accept her, we accept her! Gooble-gabble gooble-gabble, one of us!" Each of the freaks takes a sip from the goblet; the rhythm of the chanting, the excitement of the rag-tag drunken diners, and the tottering dance of the grinning dwarf on the table all sweep the scene up to a dark, hysterical pitch. When the goblet reaches Cleopatra she is repulsed. She steps back in horror and screams: "You dirty, slimy FREAKS. FREAKS! FREAKS!"

Her reaction is the nail in her coffin. As the most famous line in the film has it, "Offend one and you offend them all." The freaks plan their revenge: wearing black leather caps and looking ominous, two of the dwarfs sharpen their flick-knives and take out a gun. En route to the next town, a storm breaks out; the circus wagons get stuck in the road and topple over. Through the rain and mud we see the freaks crawling and slithering past the wheels of the carriages in an angry, choreographed assault. The Living Torso has a dagger in his mouth, which glints in a sudden flash of lightning. There is a loud scream, and the film cuts to a circus sideshow, where a barker is introducing an exhibit: "How she got that way will never be known. Believe it or not there she is." Inside a coop we see what is left of Cleopatra's face, attached to the body of a chicken.

The freaks Browning hired were already stars in their own right. Though lacking the glitz of Hollywood, they had provided popular amusement on a similar scale, and were known to anyone who had wandered along the boardwalks of Coney Island, or been to the circus that claimed to be "The Biggest Show on Earth." When Browning brought them to Hollywood, however, the reactions the film would eventually receive became evident before shooting even began.

The freaks were supposed to eat at the commissary at MGM, but too many people complained that they were putting them off their food, so a special outdoor dining area had to be set up for them. One diner remembered F. Scott Fitzgerald having a violent reaction to the Siamese twins when they sat down next to him, though that may have had more to do with something he'd drunk than with something he'd seen.

Posters for the film made it seem like prurient porn. "Can a full grown woman truly love a midget?" they asked. "What is the sex of the half man half woman? Do Siamese twins make love?" When it was released, in 1932, the most widespread reaction to the film was a basic, instinctive distaste for the freaks themselves. One woman even tried to sue the studio, claiming the film had provoked a miscarriage. MGM had anticipated this distaste, since many preview audiences had run from the cinemas. They released other ads that took a defensive tone:

A LANDMARK IN SCREEN DARING

Do we dare tell the real truth on the screen?
Do we dare hold up a mirror to nature in all its grim reality?
Do we dare produce FREAKS?

The rhetorical answer was yes; but for the studio that traded on glamour, the actual answer was—not quite. They cut the movie by a third and added a happy ending in which the midgets were reconciled. In some parts of the U.S., it exceeded all previous box office takings; in others it was a flop. In Britain, it was banned for thirty years. *Freaks* lost MGM more money than it had spent on the entire budget of *The Unholy Three*, the film that had brought Browning to prominence in the first place.

What seems odd now is that the Dolls should have been mixed up in this ruckus. In a way, the midgets weren't thought of as freaks at all, and yet they can hardly be said to be separate from the subject of the film, since they have the most prominent roles. Onscreen they are the good guys, those who have been wronged and must be avenged. Offscreen, they were not treated like the others. Daisy, Harry, and the Hilton sisters were the only members of the cast who were allowed to eat at the commissary after the complaints about loss of appetite. Leila Hyams, who plays a sympathetic big person in the film, remembered being in Browning's office with Daisy on one occasion, and noticing something odd. They were looking through casting photos of the other freaks. "Oh my," said Daisy, lingering over one of them, "it must be *dreadful* to be like that."

It was once fashionable, in circles of what now seems like pseudo-science, to try to identify whether an oddity exhibited at a fair or dime museum was actually a new species, or a *lusus naturae*. The Latin term is generally translated as "freak of nature"; but the word *lusus*, taken in this context to mean a mistake, is actually a noun derived from the verb *ludere*, "to play." So a *lusus naturae* can also be thought of as nature's toy or doll—in which case the Doll Family are, as it were, perfect "freaks," more so than any of the rest.

The Doll Family

The cultural historian Leslie Fiedler has identified what he calls "scale freaks" as the most enduring. Midgets, Giants, Fat Ladies, and Human Skeletons all function as a perspectival trick: like the magic potion that made Alice in Wonderland "shut up like a telescope" or the cake that made her stretch to more than nine feet high, these freaks skew their viewers' sense of scale. Dickens gave an evocative account of this when he saw the dwarf at Greenwich Fair:

> The best thing about a dwarf is that he always has a little box, about two feet six inches high, into which, by long practice, he can just manage to get, by doubling himself up like a boot-jack; this box is painted outside like a six-roomed house, and as the crowd see him ring a bell, or fire a pistol out of the first floor window, they verily believe it is his ordinary town residence, divided like other mansions into drawing rooms, dining parlour, and bedchambers.

As Fiedler points out, the word "monster" (the traditional name for a freak) comes from the Latin *monstrare*, to show, and *monere*, to warn. So, to those who saw them, it was as if the humans on display came with a double purpose, ready-made: they existed to be exhibited, and to warn their audiences of some ultimate moral—to show what the viewer was not, and to warn of what he or she might become.

This idea that sideshow freaks were decidedly "other" masked another thought: that they were in fact, as Fiedler has it, our "secret selves." The freak is a secret self in the same way as the android is: both elicit a similar anxiety. People who went to see the Automaton Chess Player or Vaucanson's flautist, or heard of the latter's plans to make an anatomical machine, were all faced with the question of what these "others" were that they themselves were not. Were the "objects" (as even living curiosities were thought of) safely separate from the human domain, or had they usurped some fundamental right or philosophical dimension that should only have been ours? Is the midget before us a doll, or is he or she really alive? If he or she is alive, how can he or she be so small? If the midget is in fact a machine, can this machine think? Can it talk? Can it bleed? What part of ourselves, secret or otherwise, has it taken? The android works on the viewer's mind in the same way as any curiosity—natural or artificial—in human form.

As the cultural critic Susan Stewart has put it, "the freak of nature is always a freak of culture." Freaks were not born to be on display; they were cast into the realm of curiosity by those who saw them. Just as the doll is a focus for human projection—"the gruesome foreign body," as the poet Rainer Maria Rilke put it, "on which we squandered our purest affection"—for the audience, the sideshow midget is a being more imagined than real. The human spectacle is all about possibility, and not at all about authenticity. Just as Barnum claimed that Joice Heth was 161 years old, and that the child Charles Stratton was a general named Tom Thumb, if the Schneider family called themselves dolls, then that was what audiences were expected to believe they were.

All of these shows were culturally constructed. As Susan Stewart points out, "there are no miniatures in nature": if we think of something as a miniature, rather than simply small in its own right, then we are imagining it in relation to ourselves and our world. The miniature is a construction, thought of as a shrunken replica of something else. So to imagine the Doll family as dolls—as miniatures—and to present and display them as such, is to think of them as tiny versions of ourselves: not people with their own lives, in other words, but human simulacra, just like the androids of previous centuries.

In 1938, the Doll Family were given roles as Munchkins in *The Wizard of Oz*. They were four of 122 midgets hired by Leo Singer, an impresario who had been asked by MGM to gather them all together. Singer already had his own troupe of performing midgets, and though there weren't as many of them as the studio had requested, he had some experience of assembling little people en masse. In 1913, Singer, then a producer of musical comedies in Vienna, had seen a painting called *The Wedding of the Dwarfs*. The scene depicted was one arranged by Peter the Great, who had built a miniature snow village in Russia in 1710, and summoned all the midgets and dwarfs in his empire to live there. Inspired by this historic event, Singer set out to do the same. He had a toy city built in Vienna by Joseph Urban (who would later work as a set-designer for Florenz Ziegfeld and in Hollywood), and recruited 125 midgets to live there. Thirty-five of them came with him on a tour of America, and they never left.

Singer oversaw the group of midgets he hired for *The Wizard of Oz* once they got to Hollywood. Their behaviour became the stuff of legend. In 1967, during a guest appearance on *The Jack Paar Show*, Judy Garland was asked about the Munchkins. "They were little drunks," said Garland, an actress not best known for her sobriety. "There were hundreds of them. Thousands! They put them all in one hotel in Culver

City. And they got smashed every night and [the MGM people] picked them up in butterfly nets."

Garland wasn't the only one to report such incidents. The film's producer, Mervyn LeRoy, later said that "they had sex orgies in the hotel and we had to have police on just about every floor." Bert Lahr, who played the Cowardly Lion, recalled that there were 350 midgets, busloads of whom once bared their behinds out of the windows. On other occasions, Lahr claimed, the midgets' costumes proved too cumbersome for them, and whole teams of people were required to take them to the bathroom. A story soon circulated about one of them falling into the toilet bowl and having to be "saved from drowning." An MGM wardrobe mistress said she had tried to help one of the Munchkins to get undressed, thinking he was a child. When he turned out to be a bit shy, she said, "Come on now, I have a little one just like you at home," and found, to her great surprise, that the fully adult midget wasn't like her little one at home at all.

Although not every rumour about the Munchkins was true, or certainly not true of all of them, Singer seems to have had his share of troublesome episodes. One midget, Charles Kelley, was repeatedly violent: he once walked on to the set with a couple of real guns at his side; he threatened his wife and another midget with a knife; he grabbed his wife by her hair and dragged her to the ground while everyone was having breakfast. Before long, Kelley and two other midgets, managed by a man named Powell, were fired. An MGM internal memo, requesting that they be sent home, stated that "Mr. Kelley tried to kill Mrs. Kelley last night, and one of Powell's midgets tried to knife his assistant. Pleasant little fellows."

Nevertheless, Singer evidently ensured that he was well compensated for these incidents. MGM paid each Munchkin a weekly salary of $100. None of them ever received more than fifty. Singer, allegedly, was skimming off half of their pay. Even the dog that played Toto got $124 a week.

The Doll Family, however, had little part in any of this. They were under separate contracts, having worked in Hollywood before, and, rather than staying in the same hotel as all the others, they decided to rent a place of their own. Grace, Daisy, and Tiny can all be seen in the Munchkin throng, and Harry is more prominent as one of the three Munchkins in the "Lollipop Guild."

Yet the confusion with children, and the idea of miniaturization, still would not leave them. Kids were constantly being brought on set to visit the Munchkins, and they were even honoured once with a visit from Greta Garbo, whose curiosity got the better of her when she was told there was a midget who looked just like her. She came, she looked at the little woman, and left. Whatever she may have thought went unrecorded.

Years later, a toy company in New Jersey decided to manufacture a set of "munchkin dolls." They were seven inches high, made of rubber, and one of them, based on Harry, was a member of the Lollipop Guild.

In 1927, John Ringling moved the winter quarters of his circus from Bridgeport, Connecticut, to Sarasota, Florida. The Doll Family liked it so much there that six years later they moved to Sarasota permanently from Pasadena. In 1935, they bought their first home, a five-room house on Bee Ridge Road: it had a garden, where they put a paddling pool, and was surrounded by citrus trees. Grace looked after the flowers, and often said that if it hadn't been for the circus she would have opened a flower shop. Tiny and Harry tended to the chickens in the yard and the grass—it took both their strength to push the lawnmower. Daisy continued to make clothes, and did all the cooking. The family even got a small Pomeranian dog called Jojo. (The breed is a miniature version of the Arctic Spitz, and originated in Germany in the eighteenth century. The dogs were gradually bred to be smaller and smaller until they reached their current standard miniature size.)

The light switches in the house were placed lower down than usual on the walls, and the Dolls had a set of small chairs and children's cutlery (though they were used to eating with larger knives and forks at the circus). Other than that, everything in their house was of normal size, with slight alterations made for their use—shortened table legs, for example, and a platform to stand on by the stove.

The place became known, inevitably, as "the Doll house," but the real doll's house was behind it. The family built a miniature playhouse, a white clapboard construction with a porch, in the backyard, where Tiny spent a lot of her time showing visitors around. She was proud of its electric lights and small wicker furniture, and the four Dolls were photographed on the porch once, for a Sarasota paper—they were local heroes in the circus town before long. In 1937, Harry, Grace, and Daisy became American citizens. Tiny became one two years later.

In 1937, the *New York Post* ran an article with a picture of Daisy, announcing that she was looking for a husband. "Daisy

Doll met the press over luncheon at Schrafft's, 566 5th Avenue," the article reported, "wearing a red wool dress she made for the occasion. The husband must be at least 5 foot, but not over 5 foot 3—'big enough to drive an auto'—and weigh 100–105 pounds." Daisy was not at a loss for suitors. "Dressed in her gowns she was lovely," one admirer remembered, "actually a 'living doll.' How she attracted men! And I mean full-grown men flipped over her."

Daisy soon found the man she would marry—though he was a little taller than she had bargained for. Louis Runyon was an usher at the Ringling circus. He was 5 feet 5 inches, and preferred to be called Bob Montgomery. The Dolls, with inimitable irony, called him "Pee Wee." Daisy was with Runyon for a while before they married, at the end of 1942. He moved in with the four of them, and had his uses—he drove the Dolls around, and even drove them all the way to California in 1938, when shooting began on *The Wizard of Oz*. While they worked, he sat around, looking after Jojo, smoking cigarettes, and reading the funnies; but he was a drinker and a gambler, and, it was later said, "Daisy wasn't for that." They didn't have children, and the marriage only lasted a year.

Harry also hoped he would marry some day. He told the reporter from *Collier's* magazine that he had once had a relationship with a French midget he met at a carnival, but that hadn't worked out. He imagined he would retire eventually, live in Sarasota with his future wife, and open a cigar store—he only needed to find the right midget woman, and they would raise a normal-sized family. This was at least medically possible, but of all the Dolls, only Daisy ever married.

Surrounded by his sisters, Harry was quite a patriarch in his way. He was fond of fine clothes and rich food; he enjoyed a mellow drink and a strong pipe; and, according to everyone who knew him, he loved a good joke. (This was true of all the family—as one article put it, "as kidders, there is nothing small about their sense of humour.") His habits in Sarasota were those of a keen sportsman. He liked to compare his height to that of a driving iron, and often said he weighed the same amount as "a good bird dog." He used to hunt and fish— he shot quail, doves, rabbits, and even rattlesnakes with a shot- gun adapted to suit his size. He told *Collier's* magazine that "those short barrels make the shotgun sound like a cannon . . . and boy does it kick!" Harry also had a motor boat, which he called "The Little Skipper." He and his sisters were pho- tographed on an outing in it, dressed in identical sailor suits, with Jojo tagging along.

The Doll Family retired from the circus in 1958, and later moved to an address in Water Oak Lane, when Bee Ridge Road became a busy thoroughfare; but they remained in touch with the circus community. Many of the people they had been on tour with lived nearby, and they made occasional visits to special events at the circus museum in Sarasota, hailed as returning stars.

Their deaths were reported as follows: Grace died in 1970; a hundred circus friends attended her funeral, and her ashes were strewn over the much-loved roses in her garden. Daisy died in 1980, and Harry in 1985. Harry, being the best known of the four, received the most coverage. Because he had worked in Hollywood, his obituary appeared in *Variety*, and there was a brief notice in the *Sarasota Herald Tribune:*

The Doll Family

KURT "HARRY DOLL" SCHNEIDER, 83,
>   21 *** Street, Sarasota, died Saturday (May 4, 1985) at
>   Doctor's Hospital.
> Born in Germany, he came to the area 51 years ago from
>   California.
> He was a performer with Ringling Bros. and Barnum &
>   Bailey Circus.
> He leaves three sisters and a brother.
> There will be no services.
> National Cremation Society is in charge.

Of Tiny I could find no trace. Who were the three sisters who survived Harry? Were they the ones left behind in Germany, or could one of these Dolls still be living?

For the last few decades, Sarasota has been mainly a retirement area, unfondly known as the home of "the newly wed and the nearly dead." It was once home to more ex-Munchkins than any other place in the world. Now it is booming: everywhere you look there seems to be something under construction—a flashy hotel by the beach, a concert hall, condominiums, a mall. The city is full of cinemas and restaurants, and there is an imposing new library. Sarasota is undergoing a renaissance of which John Ringling, an incurable Italophile, would have been proud. A bohemian bookshop even sells coffee in hand-painted Italian mugs.

It was Ringling who first made Sarasota the luxurious travel destination that beckoned moneyed Americans in the late twenties and thirties. He built bridges and causeways; in a multi-million-dollar development, he linked the Keys to the mainland and created a glamorous beach resort. To prepare for the arrival of the circus, he spent $500,000 on the erection of new buildings, and made sure enough railways and wagons were built or repaired. Jobs were created for and kept by hundreds of thousands of people, and Ringling's investment saved the city from an early economic crash. Chamber of Commerce ads from 1927, the year Ringling brought his circus here, refer to its many "splendid hotels" and "miles of splendid streets," claiming Sarasota as "Florida's Most Beautiful City."

Once the clown in his family's circus troupe, John Ringling bought the Barnum & Bailey circus with his four brothers in 1918. It was then that the coalition became known as "The Biggest Show on Earth." By the time he settled in Sarasota with his first wife, Mable, John Ringling had amassed quite a fortune. As his winter residence, he had an enormous palazzo built on the water, which he called Cà d'Zan ("House of John" in Venetian dialect), and he constructed a museum in the grounds for his numerous art treasures. An art school was also set up nearby. Amongst other things, the museum still contains one of the country's best collections of seventeenth-century Italian paintings, and the world's largest Rubens collection in private hands. Ringling died in 1936, at the age of seventy; he is buried in the grounds he bequeathed to the state. Both of the original buildings, and a circus museum added in 1948, are open to the public.

Cà d'Zan is a sumptuous folly—decadent and out of place, a gothic palace in the tropics. It is surrounded by feverishly sprawling banyan trees, their roots growing in mid-air like monstrous limbs. The trees were a gift to John Ringling from Thomas Edison, who also spent his winters in Florida, not far from here. (It was at his botanic garden in Fort Myers that Edison conducted his experiments with rubber.)

Though Cà d'Zan is being repaired, and much of it is cordoned off, you can wander in and out of the stained-glass atrium, past untouched bedrooms and theatrical winding stairways. A recent film adaptation of *Great Expectations*, starring Anne Bancroft as a sultry Miss Havisham, was filmed here, and you can see why: the place is full of a haunting residue, of forgotten splendour. The refrigerators and ovens take up whole rooms downstairs; there are diamond-shaped Venetian glass tiles, turrets hidden in corners, and crumbling frescoes. At the back of the building, doors open out on to a broad, tiled terrace that ends in a once-grand stone staircase. Straight ahead is Sarasota Bay: where the steps have fallen away, the sea laps against the terrace, and stretches to infinity.

Inside the art museum, a pillared white villa built around manicured Italianate gardens, huge marble fountains, and Grecian bronzes, is an extraordinary collection. I walk past a Velázquez, a Cranach, a Louis XV mirror. Yet these magnificent, wholesale imports lead up to a rather bizarre vision. At the far end of the gardens is a full-scale replica of Michelangelo's David; and behind him is a row of palm trees. This, you soon realize, is no ordinary museum: it's Old Masters in the Sunshine State.

It all seems crazy—the loot, the jungle, the palazzo on the sea—but then the tropical David imposes itself, as a kind of clue to Sarasota's quirky logic. The city achieved its glitz because the circus arrived—not from casinos or movie stars, but because clowns, acrobats, elephants, and freaks came to live here—and where the inelegance of elephants leads to glamour, nothing seems mad any more. Later, I notice that David has been carved in pink on Sarasota's road signs—he has become the emblem of the city. The casual appropriation is fantastic: it's as if Michelangelo had come here to retire.

When Harriet Beecher Stowe wrote about Florida in the nineteenth century, she remarked on its "peculiar desolate untidiness." "Florida," she wrote, "like a piece of embroidery, has two sides to it—one all tag-rag and thrums, without order or position; the other side showing flowers and arabesques and brilliant colouring." She was referring to the natural environment, but in Sarasota this double-sided quality persists even in the urban landscape. Not all of the city is like Ringling's estate, and money has not always been poured into it. The road that leads up to the museums is a highway, lined with clammy motels that the fifties left behind, and much of the city is like that—broad empty streets flanked by houses that are low, anonymous boxes—a spread-out suburbia. The Works Project Administration *Guide* to Florida, written in 1939, noted that Sarasota was "at once . . . a floodlighted stage of frivolity and a behind the scenes struggle for existence." In that respect, I find the city unchanged—it has a sort of peopled vacancy. It's not a ghost town, but it is a town off duty, no longer in its heyday—a place built for happy hour in which, as anywhere else, not all the hours can be happy.

The German actress and singer Lotte Lenya once said that when she and her fellow exiles arrived in New York from Germany in 1935, they were already familiar with the city, since they had seen it so many times in the films of Von Sternberg and Von Stroheim. It felt, she said, like coming home. With her husband, Kurt Weill, and playwright Bertolt Brecht, Lenya had already done her share of importing America to the German stage. They took lumberjacks and gangsters, gave them a minimal set and German songs, and sprinkled the whole with rugged irony. In the process, Weill and Brecht, with Lenya as their voice, turned American myths into a German invention—stylized, stylish, more bitter than ever.

Though the Dolls left Germany earlier, there was something Lenya-like about their image as circus performers, and about their screen style. They preserved strong German accents despite their increasingly idiomatic English, which turned their voices into high-pitched Brechtian gangster-speak. "Like hell you will," Harry tells someone in *Freaks*, or, in the title cards of *The Unholy Three:* "If you tip that boob off to who we are I'll put some lilies under your chin." Harry and his sisters looked and sounded as though they might have been cut out more for a life of sophisticated crime than for a life at the circus.

Was there any of Lenya's sentiment in the Dolls' decision to settle in Florida? Was it so much the epitome of luxurious America as they saw it, such a familiar image of the good life, that coming to Sarasota felt like coming home? In my mind I see the Dolls, dressed in their Coney Island finery and strolling along the boardwalk, past the motor boats, amidst breezy palms and Florida pinks. They are a distinguished-looking bunch, tiny old-timers who've lost their way to the cabaret. I can't help feeling that something about the place must have made them feel comfortably un-strange, that they came here prepared to live out the kind of American Dream that might have visited the nights of Angela Carter.

After trawling through old photographs and newspapers

in the circus archives, wearing white gloves and listening to the calliope playing in the museum next door, I decide to pay a visit to the address printed with Harry's obituary. *** Street is in a serene-looking housing estate patrolled by a security guard. There is no one in the booth, so I drive on, slightly taken aback by this private, slumbering paradise with its fenced-in tennis courts and canopied country club. The houses are mobile homes, which look as though they might have landed here, spinning like Dorothy's tornado-struck cabin in *The Wizard of Oz*. ("Like the snail, they carry their home around with them," a Ringling Brothers programme had said of circus people.)

There is no one around. I turn the corner, and the first house I come to is number 21. It is a sleek, mid-grey affair, with a strip of window around the middle and a screen-covered porch to the side. I am admiring the house's toned-down, rather European design, when the metal mailbox bearing its owner's name comes into view. In white capital letters stuck on to a piece of black plastic are the words "SCHNEIDER/DOLL."

I saw now that I should not be here—I had had no permission or advice. I had come, kidding myself that perhaps I could speak to a later owner—that Tiny would no longer be here, and that the house was only a clue, a way of reading backwards into the lives of those who are gone.

I was already thrown by the mailbox itself. "Doll," it made clear, was not just a stage name. It was, almost half a century after these performers retired, the name they used as their last. "Doll" had gone from being a description to being an identity, taken on whole. As I trespassed in seeming slow-mo across the grass and raised my fist to knock, I fought the idea that, despite the equanimity of my research, I might not, in real life, know how to handle a meeting with a living doll.

I knocked twice and, relieved there was no answer, walked away; but a lamp had been turned off and the television was still on. Someone was in there and did not want to be disturbed. I moved back towards the door for one last guilty attempt (I had come so far), and a voice called out, as if from nowhere, "Hey you! What do you want!"

It was the angry, German-inflected, high-pitched voice I knew from *Freaks*. Was Harry not really dead? I looked around.

"Over here!" the voice shouted.

I stepped back from the house and saw a face in the window.

"Who are you? What do you want?"

The glass had been slid open and I found myself speaking to a figure through a screen. I said I just wanted to introduce myself. I said I had been reading about her family and was an admirer of their act. I had come (I stumbled) to make a sort of pilgrimage to their house. The figure seemed relieved.

"Oh. I see. It's just that I never use that door. I always use the one at the side, so I didn't know who it could be. You live here in Sarasota?"

Her face was wrinkled, not into fine dry lines, but in broad folds of taut, youthful skin. Her small blue eyes seemed to be barely lidded, or missing lashes, or something that made them distant, set them back into her face. She had short, mid-brown hair cut in a sporty style. It was impossible to tell how tall she was, since I was standing below floor level on the grass, talking up to her, but she was certainly not one of the larger German siblings. She seemed small and trim and fit in her baggy white T-shirt. She didn't look like an old woman, and if this was the youngest of the Schneider sisters, she had to be in her mid-eighties.

"You are from England?" she asked. "I can tell. You see, I am from Germany originally."

"Yes," I said. "I know."

We were still speaking through the screen window, and she clearly wasn't inclined to let me in; but it hardly seemed to matter because by now—whether from the shock of the encounter, or the sting of her life remembered—I was moved almost to tears.

"Anyway," I muttered, stepping back and trying to regain my composure, "I just wanted to introduce myself."

She smiled. "Nice to meet you. I never use that door, you see." She repeated this several times, apologetically.

I waved goodbye and asked, as an afterthought, her name.

"Tiny," she said, "Tiny Doll. I'm sure I will see you again."

When I drove out on to the road, the security guard was back in his booth. He must have just slipped out for coffee.

Some years later, I returned to visit Tiny. I had tried writing to her, and had received no reply; but eventually one of my attempts yielded an e-mail, from her neighbour and friend Marlene Townsend Grunewald. Marlene explained that Tiny was legally blind and unable to read her own mail; she told me that she and her husband looked after her; that Tiny had had three serious eye operations but had no income and no medical insurance ("in her day midgets couldn't get it"); and that she had "been bothered by many who only are curious and who treat her like a child."

They were understandably suspicious. As Tiny later put it, in the tough speech she must have learnt from the movies: "Many people over the telephone say, 'Oh I want your story.' So I ask, 'Well, what's in it?' Then they say, 'It's for publicity.' I tell 'em, 'I'm sorry, I can't eat publicity.' " But over time, Marlene was extremely generous to me, and she arranged for me to visit Tiny one hot day in August.

Tiny's porch, I noticed this time, was lined with miniature chairs, including four little director's chairs with the Dolls' names on them, which they brought back from the set of *The Wizard of Oz*. She was there, waiting, when I arrived, and led me into her sitting room. There were two normal-sized sofas with a wooden coffee table between them, and, in front of the table, facing the television, was a small grey armchair, which must have been where she usually sat. Tiny seemed nervous, busy-bee-ish; she moved quickly, with great ease, and she appeared, that day, both impishly sturdy and somehow uncertain. She had a strong German accent, with rasping "r"s, and a very American turn of phrase—a speech Brecht couldn't have bettered. She was eighty-six years old and 39 inches tall.

"Have you ever been in a mobile home?" she asked.

I said I didn't think I had.

"Oh! Let me show you around!"

She took me through her well-equipped kitchen (it had cupboards above the sink that she must never reach), her dining area, her tidy bathroom. There was wooden panelling on the walls, and air-conditioning throughout. The ceilings felt low, but no lower than any other home of similar design. Harry's old bedroom was next to the bathroom. It had a single bed in it, neatly made. Further back was another bedroom, with an ensuite shower room that Tiny seemed to use as storage. There were twin beds here—one where Tiny slept, and another that used to be Daisy's. There were three toy dolls lined up on one of the beds, wearing crocheted outfits. "Marlene made those," Tiny said, and then, looking up all of a sudden, "You know, you're quite tall! . . . Five foot six?"

"I think it's the shoes," I said, realizing my soles were inappropriately thick. "You want to try them?"

"Ha ha ha!" She burst into rasping, hearty laughter.

"When did you move here?" I asked.

"Well," she said, thinking it over, "Grace passed away in '70, in the other house. But Daisy and Harry and myself moved in here in '80. Daisy was here three months, and then she passed away. They both had murmuring hearts. Even though they took pills. But you know, when your time comes, it comes."

"That must have been hard for you," I said.

"Well," she announced loudly, as if banishing the thought, "you gotta get used to it. Chin up and don't give up—that's for sure."

"So you've been on your own here for a while."

"Oh yeah. You get used to it. Gosh yes. For a while you're a little lonely, but then you say, oh well, you gotta do it. See, there was another midget, Nita Krebs, when she passed away she was I think eighty-seven. She was one of the ballerinas in *The Wizard of Oz*. And she lived by herself, pretty much, till she got sick, then somebody else took care of her till she passed away."

"Did you know many other little people?"

"Oh yeah, quite a few. But they're all gone."

"Your house is very spacious."

"Oh yeah," she said, proudly, and pulled out the coffee table so that we could sit down on the sofa.

It was only when we were sitting down that I noticed her ears. They seemed to have grown at the rate of a tall person's, and so were out of proportion with the rest of her face. Her light brown hair was cut short, with wisps around them. She was wearing clothes she had inherited from her brother—children's clothes—a baggy tennis shirt, dark blue cotton shorts, and Mickey Mouse trainers with Velcro straps. She sat next to me, with one leg tucked under her and the other dangling over the edge of the sofa; from time to time she would rub her knee, or fiddle with her shoe, or jump up to find an envelope or book whose title she couldn't read. We shuffled into a conversation—with me asking broad, soft questions and waiting for her to reminisce, or trying to check specific facts. Tiny would shrug off her answers, as if nothing were really important any more, and now and then she would burst into loud, chesty laughter.

"What was life like for you," I asked, "before you came to America?"

"Well, I went to school, and well, when you're young you don't think nuttin' of it."

"When did you realize you were small?"

"Ha! Don't ask me! I think you get to a certain age and

that's it. Well, Grace was about my size, Harry was a little smaller, and Daisy was the tallest little one." She got up to show me with her hand how tall they were in relation to her.

"You had some taller siblings too, didn't you?"

"Yeah, there was nine of us. Four little ones and the rest was big ones. I was the last one."

"Did you get on with your other sisters and brother?"

"OK . . . Well, when we came over we only seen them every five years, and then after that, when Hitler got in, that was out. Then after '44, my mother died, then my father died in '45. But we was in the circus, and we couldn't leave . . . The circus closed up in '56. Then we stayed home—we were gettin' elderly too. So we just rested up, because circus life is pretty tough. We started in New York, then Boston, then Philadelphia . . . It was . . . I liked it—circus was my life. In fact, we all liked it."

I asked her how she learned English.

"Some fellow in the circus, he couldn't speak German— he said, 'Can I learn Tiny how to speak American?' Harry and Grace said, 'Go ahead!' So whenever we had time, he sat me down and showed me, he'd point to things and say 'Sofa,' 'Chair' . . . just about for half an hour, and that's it. So the next day, he says, 'What's that?' And I'd say 'Chair,' 'Table,' 'Sofa' . . . well, they didn't have no sofa, but . . . that's how I learnt. And in the old-time movies, you learn by the screen." She learnt to read, in other words, by the title cards in silent films, as her speech sometimes suggests. "And Daisy learnt me how to write. Well, I'm not a perfect writer. Harry couldn't write too much German, but Grace and Daisy could." Later, on the back of a photograph of the Dolls with silent comic Harold Lloyd, I found, in her handwriting, an exact transcription of her accent: "Herald Lloyd."

Tiny described the sideshow as "a fat lady, and a skinny, and a bunch o' Hawaiians." There was no detail in her account, and I wondered if she and the other performers had been close.

"Oh God, yes," she said. "You're from morning till night with them all the time. We see 'em every day, every hour. We had all kinds of good friends. But in the circus," she added, giving a clue, perhaps, to her current tactics of detachment, "you really don't get too involved with people, too much."

Equally, when I asked about the Doll family's act, she shrugged, and said that generally, they didn't sing any "love songs, and no . . . what do you call that . . . jitterbox—no, we had to sing decent songs, because people come with their kids." I asked if she remembered anything in particular and she sighed, "Oh . . . gosh. I kinda forgot now what we all sang. Sorry."

Her eyes seemed rather distant, as I'd remembered; perhaps because she was blind. I asked her how much she could see. She said she could see "fine" from a distance, but she couldn't see anything close-up.

Everything, the way Tiny tells it, is "fine," or elicits the response "oh yeah," with the sort of soothing lilt you might use when comforting a child. She seems uneasy talking about herself, and unwilling to dwell on the past, though her memory, as she reveals in certain details, is excellent. Every suggestion of difficulty is brushed off with a sort of cliché-packed stoicism ("chin up and don't give up"; "you gotta do it"; "you get used to it"). "Never once have I heard her complain about her size," Marlene had said, "or any inconvenience she experiences because of it."

There was one moment, however, when Tiny seemed more than just edgily dismissive—she seemed genuinely troubled. I was trying to find out how difficult performances had been for her, what it was like to be on show; but I only received a by now familiar reply: "You get used to it." I asked if audiences were unpleasant to them and she replied very quickly that, on the contrary, "they were very nice people." "Some of them," she went on, "you know, they were terrible, but you don't pay any attention to them." I asked if they were treated like dolls, or children, and Tiny continued to brush off the difficulty. "Some people do," she said, "but you don't pay any attention. You get used to it."

"In *The Unholy Three*," I persisted, "Harry plays a character who really dislikes the children in the audience."

"Oh no," Tiny replied, springing rather surprisingly to her brother's defence, "Lon Chaney was the real crook. He went as a woman. And Harry was the one that stole the jewels from the rich people. And this one little girl, she thought he was one of the burglars, and well, of course it's a movie, he looked like he was trying to choke her. Well, see, Harry had to do what the producer and director tells him. He didn't want to do it."

Her reaction, I thought, was quite extraordinary, and it took me a while to understand that she was referring to the scene that had been cut—a scene I hadn't mentioned, and one of which anyone who had only seen the movie (and not its outtakes) would be unaware. I felt strangely moved by this residue of vicarious guilt, three-quarters of a century after the fact; by the idea that harm done to a child by her brother, even in a movie, and even on the cutting-room floor, should have stayed with her for so long.

"I think they took that scene out," I said, by way of appeasement.

"I think they did too," she said. Her memory was impeccable.

A little later on I persuaded Tiny to show me some of her photo albums, which she heaved out of a large chest in the living room. We went through pages and pages of well-preserved black-and-white pictures from their days in Hollywood and elsewhere: the Doll family shaking hands with Burt Lancaster; the Doll Family with Jackie Coogan, the child actor just their size; the Dolls posing either side of Harold Lloyd; with Dinah Shore; Ramon Novarro. There were photos of Daisy in various glamorous costumes, and signed portraits of other stars: Leila Hyams (one of the actresses in *Freaks*), Joan Crawford, Clark Gable, Greta Garbo. "Boy, she was a great actress," Tiny sighed over Garbo, "she was really wonderful . . . she was *tall*." I don't know quite what I was expecting from my visit to Tiny's house. I had, perhaps, wanted it to be the day a Doll finally spoke. Yet, welcoming as she was, Tiny had little to say to me. She was a player in, rather than a commentator on, her life, and it was a life that had already been subject to prying eyes and practical difficulties. She didn't really want to talk about it.

What she showed, in other words, was that, in a crucial sense, she is only human. She is a person with emotions, and rights. Unlike all the inventions in this book, which can be spo-

EDISON'S EVE

ken of and reflected upon as objects, Tiny is no doll at all. While the others approached humanity and blurred the boundary between what was living and what was not, she makes that boundary abundantly clear. By her very presence, Tiny shows that "the Doll family" was a terrible—possibly tragic—category mistake. Though she skirted these issues in our conversation, they were there in every nervous gesture and apparently indifferent word.

Before I left, I wanted to get a sense of what the Schneiders were like as a family, of the different personalities of the four who had left Germany and spent all of their lives together. When I asked Tiny, she thought for a bit, then jumped up from the sofa and crossed to the other side of the room, where, in a glass-fronted cabinet, were several rows of little trinkets. On the top shelf were porcelain tankards about two inches high, complete with pewter lids and German inscriptions in blue Gothic script. This, Tiny explained, was Harry's collection. On the next shelf down were little painted pottery figurines, tiny Hansels and Gretels holding brooms or baskets of flowers, pieces of folk kitsch. They belonged to Grace. Daisy's shelf was more miscellaneous—a ballerina in a pink tutu—dressed as Daisy herself had been in *Freaks*—and some baby deer with white spots on their backs. "They brought these back from Germany," Tiny explained. "But this is what I collect." She turned round and gestured with her arm towards a shelf near the television.

There was an army of the tiniest elephants: elephants of ivory, glass, clay, and quartz—masses of them, all lined up. "Did you like the elephants when you were in the circus?" I asked, trying to imagine the smallest person with the largest

262

animal. "Oh yeah," Tiny said, "they're my favourite animal." She explained that when they held their trunks up, that meant they were happy—and I noticed all of hers had their trunks curled that way. They were both souvenirs and transformations, so that she could remember her life at the circus and reign over it at once. In a reverse Promethean gesture, Tiny had turned her favourite animals to stone, rendered them inanimate, and shrunk them so that she could possess them. She had done to the elephants what, throughout her life, people had in their imaginations done to her.

The cabinets in Tiny's living room resembled nothing so much as old cabinets of curiosities. Just like Charles Wilson Peale, whose museum housed the Automaton Chess Player alongside prehistoric skeletons; like Gottfried Beireis, who put Vaucanson's duck on display next to stuffed and rotting birds; like the early exhibitions in the Egyptian Hall or the far-fetched contents of Thomas Edison's machine shop; like the botanists and classifiers of Louis XV's Jardins du Roi, Tiny had contained and collected her world around her, put the universe within reach.

I wondered then if this puzzling scene, this uncanny foxing of life and scale, was where the philosophical toy had ended up—in the Florida mobile home of a retired circus performer and movie star—and whether a living doll had not, in the end, had the last, raucous laugh.

Epilogue

The late Ichiro Kato, father of Japanese humanoid robotics, wanted to complete one last project before he died. His favourite book was *The Eve of the Future,* by Villiers de l'Isle-Adam, and all his working life he had wanted to build a perfect woman, his own "Hadaly." He made Hadaly 1, then Hadaly 2—a huge metal talking torso on wheels, with big round eyes, shoulders as broad as a football player's, and enormous black spidery fingers. But neither Hadaly was perfect enough. Kato was in his sixties, and working on another robot, when he died of a heart attack, in 1994. Hadaly was the last android he ever made.

Kato's disciple, Atsuo Takanishi, has now set up his own robotics lab at Waseda University in Tokyo. This is the place Cynthia Breazeal at MIT suggested I visit, where the robotic descendant of Vaucanson's Flute Player was constructed.

Epilogue

When I arrive, I find more echoes of the old android-makers than I could possibly have imagined. Hadaly is out of order, but others no less evocative are there. Takanishi's colleague, Hideaki Takanobu, introduces me to the lab's uncanny inhabitants: a chewing machine, made out of a real human skull and re-animated using pieces of wire Takanobu refers to as "artificial muscles"; Wabian, a gigantic walking machine that dances—a hulking reincarnation of Hoffmann's Olympia; a speaking machine, exactly like Kempelen's, operated electronically; a sense machine, which wakes up if it smells ammonia, flinches if you touch its eyelashes, follows a bright light with its eyes, and turns in the direction of loud noise—to all intents and purposes, a twenty-first-century version of Condillac's philosophical statue.

By the end of our tour, Takanishi's flute-player is ready to perform. On top of a set of metal shelves, wired up to various computers, is a plastic cylinder with a piston inside it, mimicking the lung capacity of an adult man. Above that is a white mask and a black top hat, behind which are moveable, elasticated lips and a flexible tongue. To one side there is a flute, held by metal fingers, whose tips are coated in rubber.

While the machine is being prepared, I ask Professor Takanishi if he constructed it in tribute to Vaucanson. He says not at all. In fact, he doesn't believe Vaucanson's Flute Player can ever have existed, because, he tells me, "a flute-player is very difficult to make." And then, as if to contradict him—as if, indeed, it had a mind of its own—the artificial musician begins to play a song, seemingly about its ancestors: "Yesterday . . ."

Later, as I wander through the streets, dizzy from these disarming relics of the mechanical past, I think how fitting it is that this story, which began in seventeenth-century France, should end here, in twenty-first-century Tokyo, the city so loved by science fiction. There is a virtual Paris here: replicas of the Café de Flore, of Angelina's famous chocolate house, and of the domed Printemps department store. The landscape is that of *Blade Runner* and *Neuromancer:* impossibly modern, with its teeming streets, its bright lights and skyscrapers, and yet somehow post-apocalyptic, over the edge. In winding alleys reminiscent of a war, older lipsticked ladies sit in tiny rooms, hosting speakeasies past their prime. At regimented street crossings, people move together like automata ("What do I see," Descartes wrote of the passersby outside his window, "but hats and coats that cover ghosts or simulated humans, which move only by springs?"). At every convenience store there is sex for sale, and advertisements for call girls of indeterminate sex. Beneath a neon palm tree, to the tinkling sound of coins falling mechanically in the pachinko parlours, a man slips under a plastic curtain into the steam of a noodle stand, a déjà vu of Harrison Ford in his search for replicants.

The city is real, but it contains other worlds, left over from the past, or imagined from the future: science fiction has mutated into urban fact. And in this place, the passions that drove Vaucanson and Villiers, Condillac and Hoffmann, are still alive, or born again. An android plays the flute; a metallic statue is awoken by a sound or a smell; Hadaly, the perfect woman, is in disrepair.

Acknowledgements

I have received a great deal of help in the course of writing this book; many people have given their assistance with a degree of goodwill I could never have expected, and for which I'm extremely grateful. I'd like to thank Rodney Brooks, Cynthia Breazeal, Brian Scassellati, and Bryan Adams at the Artificial Intelligence Lab at MIT; Caroline Junierclerc, Marie-Josée Golles, and Anne de Tribolet at the Musée d'Art et d'Histoire in Neuchâtel; Alain Mercier, Fredérique Desvergnes, and Isabelle Taillebourg at the Conservatoire National des Arts et Métiers in Paris; Doug Tarr and Leonard De Graaf at the archives of the Edison National Historic Site in West Orange, New Jersey; Charles Silver at the Film Studies Center at the Museum of Modern Art in New York, and Helena Robinson at the Film Stills Archive there; Madeleine Mathête-Méliès, for her encouragement in relation to the work of her late grand-father; Laure Bouissou and Gaëlle Vidalie at the Cinéma-thèque Française; Daniel Fromont, Pierrette Lemoigne, Eric Le Roy, and Michelle Aubert at the archives of the Centre National de la Cinématographie in Bois d'Arcy; Kathleen Dickson and Steve Tollervey at the British Film Institute;

Acknowledgements

Kevin Brownlow and Lynne Wake at Photoplay Productions in London; Bernice Zimmer at Baraboo Circus World Museum in Wisconsin; Deborah Walk at the Ringling Circus Archives in Sarasota; and Atsuo Takanishi and Hideaki Takanobu at the Takanishi Lab at Waseda University in Tokyo. I'd also like to thank the staff of the British Library, the Bibliothèque Nationale in Paris, the Library Company of Philadelphia, the Pennsylvania Historical Society, the New York Public Library, and Princeton University Firestone Library.

A substantial part of this book could not have been written without Elly Schneider (Tiny Doll), and her neighbour and friend, Marlene Townsend Grunewald. I am grateful to both of them for the exceptional kindness they showed me.

For different reasons, I am indebted to the Master and Fellows of St. John's College, Cambridge, whose Harper-Wood Studentship enabled me to finish this book and make headway with another. Lisa O'Kelly, Jane Ferguson, Roger Alton, John Mulholland, and other colleagues at the *Observer* indulged my mad mutterings, and gave me time off to write.

Over the past few years, through a number of conversations, many people have directed my thoughts. They may not recognize their influence on the finished product, but they have either been responsible for its beginnings, or had an effect on its evolution. These people include Tim Adams, Tom Campbell, Paul Hammond, Colin Jones, John Kerrigan, Adam Levy, Andrew O'Hagan, Marina Warner. My editors—Julian Loose at Faber and Robin Desser at Knopf—and my agent, Derek Johns, will, I hope, recognize their hand, since this book is what they have made it. They have been unfailingly patient,

Acknowledgements

lucid, and encouraging. One could not hope to be read more attentively; any mistakes, however, are entirely my own.

I owe a more personal debt to Mary-Kay Wilmers. Her generosity is surpassed only by that of Elena Uribe and Michael Wood, to whom this book is dedicated. Not least, I'd like to thank Chris Turner, for many ideas masquerading here as mine, but also, for more than he knows.

Finished 12/15/2007

Bibliography

INTRODUCTION

Bailly, Christian. *Automata: The Golden Age (1848–1914)*. London, 1987.

Beaune, J.-C. "The Classical Age of the Automaton," in *Fragments for a History of the Human Body*, part 1. New York, 1989.

Bedini, Silvio. "The Role of Automata in the History of Technology," *Technology and Culture* 5, no. 1 (Winter 1964).

Boehn, Max von. *Puppets and Automata*. New York, 1972.

Capek, Karel. *Rossum's Universal Robots*. London, 1923.

Caudill, Maureen. *In Our Own Image*. Oxford, 1992.

Cohen, John. *Human Robots in Myth and Science*. London, 1966.

Collodi, Carlo. *Pinocchio: The Story of a Puppet*. New York, 1932.

Cooke, Conrad William. *Automata Old and New*. London, 1893.

De Solla Price, Derek J. "Automata and the Origins of Mechanism and Mechanistic Philosophy," *Technology and Culture* 5, no. 1.

Eco, Umberto, and G. B. Zorzoli. *A Pictorial History of Inventions*. London, 1961.

Florescu, Radu. *In Search of Frankenstein*. London, 1996.

Freud, Sigmund. "The Uncanny," in *The Penguin Freud Library*, vol. 14. London, 1990.

Godwin, William. *Lives of the Necromancers*. London, 1834.

Helmholtz, H. L. F. von. *Popular Lectures on Scientific Subjects*. London, 1873–81.

Hesiod. *Theogony* and *Works and Days*. London, 1973.

Hillier, Mary. *Automata and Mechanical Toys*. London, 1988.

Hodges, Andrew. *Turing*. London, 1997.

⦿Hoffmann, E. T. A. "The Sandman," in *Tales of Hoffmann*. London, 1982. *and The Automada*

Huyssen, Andreas. *After the Great Divide*. London, 1986.

Jones, Frederick L., ed. *Mary Shelley's Journal*. Norman, Okla., 1947.

Menzel, Peter, and Faith D'Aluisio. *Robo Sapiens*. Cambridge, Mass., 2000.

MIT Artificial Intelligence Laboratory. *Research Abstracts*. Cambridge, Mass., 2000.

Paracelsus. *On the Nature of Things*. London, 1850.

Penrose, Roger. *The Emperor's New Mind*. Oxford, 1989.

Praz, Mario. *Three Gothic Novels*. London, 1968.

Rousseau, Jean Jacques. *Emile*, tr. Barbara Foxley. London, 1911.

Schwartz, Hillel. *The Culture of the Copy*. New York, 1996.

Shelley, Mary. *Frankenstein, or The Modern Prometheus*. London, 1818.

Bibliography

Shelley, Mary, and P. B. Shelley. *History of a Six Weeks' Tour.* London, 1817.

Warner, Marina. *Monuments and Maidens: The Allegory of the Female Form.* London, 1985.

CHAPTER ONE

"Précis de la vie de M. le Cat," in *Mémoires de Trévous,* November 1768.

Altick, Richard. *The Shows of London.* Cambridge, Mass., 1978.

Antoine, Michel. *Louis XV.* Paris, 1989.

Baillet, Adrien. *Vie de Monsieur Descartes.* Paris, 1992.

Barthélemy, Guy. *Les Jardiniers du Roi.* Paris, 1979.

Biographie universelle, ancienne et moderne. Paris, 1811–53.

Brewster, David. *Letters on Natural Magic.* London, 1882.

Carr, J. L. "Pygmalion and the *Philosophes,*" *Journal of the Warburg and Courtauld Institutes* 23 (1960).

Chapel, Edmond. *Le Caoutchouc et la Gutta-Percha.* Paris, 1892.

Chapuis, Alfred. *Les Automates dans les oeuvres de l'imagination.* Neuchâtel, 1947.

Chapuis, Alfred, and Edmond Droz. *Les Automates.* Neuchâtel, 1949.

Chapuis, Alfred, and Edouard Gélis. *Le Monde des Automates.* Paris, 1928.

Chasseloup-Laubat, F. de. *Francois Fresneau, . . . Père du caoutchouc.* Paris, 1942.

Chovet, Abraham. *An Exploration of the Figure of Anatomy, wherein the circulation of the blood is visible, through glass veins and arteries . . .* London, 1737.

Bibliography

Cole, F. J. "A History of Anatomical Museums," in *A Miscellany Presented to J. M. Mackay*. London, 1914.

———. *A History of Comparative Anatomy*. London, 1944.

Coleby, J. *The Chemical Studies of J. B. Macquer*. London, 1938.

Condillac, Etienne Bonnot de. *Traité des sensations*. London and Paris, 1754.

Condorcet. "Eloge de Vaucanson," in *Oeuvres II*. Paris, 1847.

Cooke, Lynne, and Peter Wollen, eds. *Visual Display*. Seattle, 1995.

Darnton, Robert. *Mesmerism and the End of the Enlightenment in France*. Cambridge, Mass., 1968.

Decremps, Henri. *La Magie blanche dévoilée*. Paris, 1784.

Delaunay, P. *Le Monde médical Parisien au dix-huitième siècle*. Paris, 1906.

Descartes, René. *Les Passions de l'ame*. Paris, 1726.

———. *Traité de l'homme*. Paris, 1729.

———. *"The World" and Other Writings*, ed. Stephen Gaukroger. Cambridge, Eng., 1998.

Diderot, Denis, and Jean Le Rond d'Alembert, eds. *Encyclopédie, ou, Dictionnaire raisonné des sciences, des arts et des métiers*. Geneva, Paris, and Neuchâtel, 1754–72.

Doyon, André, and Lucien Liaigre. *Jacques de Vaucanson, mécanicien de genie*. Paris, 1966.

Exhibitions of Mechanical and Other Ingenuity. A scrapbook of advertisements, ranging 1700–1840.

Fayol, Amédée. *Le Caoutchouc*. Paris, 1909.

Flourens, P. *Histoire de la découverte de la circulation du sang*. Paris, 1854.

Gaxotte, Pierre. *Louis XV*. Paris, 1980.

Bibliography

Gelfand, Toby. *Professionalizing Modern Medicine.* Westport, Conn., 1980.

Goethe, J. W. von. "Annals, or, Day and Year Papers," in *The Autobiography of Goethe,* vol. 2. London, 1901.

Gonon, P. M. *Vaucanson à Lyon en 1744.* Lyon, 1844.

Gunderson, Keith. *Mentality and Machines.* London, 1985.

Haviland, Thomas N., and Lawrence Charles Parish. "A Brief Account of the Use of Wax Models in the Study of Medicine," *Journal of the History of Medicine,* January 1970.

Hewett-Thayer, Harvey W. *Hoffmann: Author of the Tales.* Princeton, N.J., 1948.

Hossard, Jean. *C. N. Le Cat: 1700–1768.* Rouen, 1968.

Isherwood, Robert M. *Farce and Fantasy: Popular Entertainment in 18th Century Paris.* New York, 1986.

Jacques Vaucanson, catalogue from the Musée National des Techniques, Paris, 1983.

Jones, Colin, and Laurence Brockliss. *The Medical World of Early Modern France.* Oxford, 1997.

Jordanova, Ludmilla. *Sexual Visions: Images of Gender in Science and Medicine between the 18th and 20th Centuries.* New York, 1989.

La Mettrie, Julian Offroy de. *L'Homme machine.* Paris, 1981.

————. *Man a Machine,* tr. Gertrude Carman Bussey. La Salle, Ill., 1961.

La Morinerie, Baron de. *Les Origines du caoutchouc.* La Rochelle, 1893.

La Suisse Horlogère, international edition, no. 3, 1949, letter from Hofrath Beireis, sent from Helmstadt in 1785.

Le Sueur, Achille. *La Condamine.* Amiens, 1910.

Bibliography

Leslie, Anita, and Pauline Chapman. *Madame Tussaud, Waxworker Extraordinary.* London, 1978.

Lysons, Daniel. *Collectanea; or, a collection of advertisements and paragraphs from the newspapers . . . ,* vol. 2, ii. Strawberry Hill, Eng., 1826.

McManners, John. *Death and the Enlightenment.* Oxford, 1981.

Paul, Charles B. *Science and Immortality: The Eloges of the Paris Academy of Sciences (1699–1791).* Berkeley, Calif., 1980.

Pyke, E. J. *A Biographical Dictionary of Wax Modelers.* Oxford, 1973.

Quesnay, François. *Observations sur les effets de la saignée.* Paris, 1730.

Raggio, Olga. "The Myth of Prometheus: Its Survival and Metamorphosis up to the 18th Century," *Journal of the Warburg and Courtauld Institutes* 21 (1958).

Robert-Houdin, Jean-Eugène. *Memoirs of Robert-Houdin,* tr. Lascelles Wraxall. New York, 1964.

Rodis-Lewis, Geneviève. *Descartes: His Life and Thought,* tr. Jane Marie Todd. Ithaca, N.Y., 1998.

Rosenfeld, Leonora Cohen. *From Beast-machine to Man-machine.* Oxford, 1941.

Sablière, Jean. *De l'Automate à l'automatisation.* Paris, 1966.

Schaffer, Simon. "Enlightened Automata" in Clark, Golinski, and Schaffer, eds., *The Sciences in Enlightened Europe.* Chicago, 1999.

Sewell, William H. "Visions of Labor," in Kaplan and Koepp, eds., *Work in France.* Ithaca, N.Y., 1986.

Stafford, Barbara Maria. *Artful Science.* Cambridge, Mass., 1994.

———. *Body Criticism.* Cambridge, Mass., 1991.

Bibliography

Tiffany, Daniel. *Toy Medium: Materialism and Modern Lyric.* Berkeley, Calif., 2000.

Vartanian, Aram. *Diderot and Descartes: A Study of Scientific Naturalism in the Enlightenment.* Princeton, N.J., 1953.

———. *La Mettrie's L'Homme Machine: A Study in the Origins of an Idea.* Princeton, N.J., 1960.

Vaucanson, Jacques de. *Le Mécanisme de fluteur automate.* Paris, 1739.

Wellman, Kathleen. *La Mettrie: Medicine, Philosophy and Enlightenment.* London, 1992.

CHAPTER TWO

"Automate Joueur d'Echecs," *Magazine Pittoresque,* vol. 2, 1834.

"La Vie et les aventures de l'automate joueur d'échecs," in *Le Palamède,* 1836.

"L'Automate joueur d'échecs," in *Le Palamède,* 1839.

Allen, George, in D. W. Fiske, *The Book of the First American Chess Congress.* New York, 1859.

———. Letters in the collection of the Library Company of Philadelphia.

Altick, Richard. *The Shows of London.* Cambridge, Mass., 1978.

An Oxford Graduate. *Observations on the Automaton Chess Player.* London, 1819.

Benjamin, Walter. *Illuminations.* New York, 1968.

Binet, Alfred. *Psychologie des grand calculateurs et joueurs d'echecs.* Paris, 1894.

Bradford, Gamaliel. *The History and Analysis of the Supposed Automaton Chess Player.* Boston, 1826.

Bibliography

Brewster, David. *Letters on Natural Magic.* London, 1882.

Carroll, Charles Michael. *The Great Chess Automaton.* New York, 1975.

Catalogue of the Chess Collection of the late George Allen Esq. Philadelphia, 1878.

Chapuis, Alfred, and Edouard Gelis. *Le Monde des automates.* Paris, 1928.

Cockburn, Alexander. *Idle Passion: Chess and the Dance of Death.* London, 1975.

Curiosities for the Ingenious. Philadelphia, 1822.

Decremps, Henri. *La Magie blanche devoilée.* Paris, 1784.

Franklin, Benjamin. "The Morals of Chess," in *Chess Made Easy.* London, 1797.

Gosse, Philip. *Dr. Viper: The Querulous Life of Philip Thicknesse.* London, 1952.

Hankins, T., and R. Silverman. *Instruments and the Imagination.* Princeton, N.J., 1995.

Hoffmann, E. T. A. "Automata," in *The Best Tales of Hoffmann.* New York, 1967.

Hooper and Whyld, eds. *Oxford Companion to Chess.* Oxford, 1984.

Hunneman. *A selection of Fifty Games from Those played by the Automaton Chess Player . . . In London.* London, 1820.

Hutton, Charles. *Mathematical and Philosophical Dictionary.* London, 1796.

Jones, Ernest. "The Problem of Paul Morphy: A Contribution to the Psychology of Chess," in *Essays in Applied Psychoanalysis,* vol. 1. London, 1951.

Mouret, J.-F. *Traité élémentaire du jeu d'échecs.* Paris, 1836.

Nabokov, Vladimir. *The Luzhin Defense.* London, 1994.

Bibliography

Poe, E. A. "Maelzel's Chess Player," in *The Complete Tales and Poems of Edgar Allan Poe*. New York, 1982.

Racknitz, Joseph Friedrich Freiherr zu. *Über den Schachspieler des Herrn von Kempelen*. Leipzig, 1789.

Robert-Houdin, Jean-Eugène. *Memoirs of Robert-Houdin*, tr. Lascelles Wraxall. New York, 1964.

Saint-Amant. "Note on Automaton Chess Player," *Le Palamède*, vol. 1, no. 5, 1842.

Schaffer, Simon. "Babbage's Dancer and the Impresarios of Mechanism," in Spufford and Uglow, eds., *Cultural Babbage: Technology, Time and Invention*. London, 1996.

Sellers, Charles Coleman. *Mr. Peale's Museum: Charles Wilson Peale and the First Popular Museum of Natural Science and Art*. New York, 1980.

Thicknesse, Philip. *The Speaking Figure and the Automaton Chess Player Exposed and Detected*. London, 1784.

Walker, George. *Chess and Chess-Players: Consisting of Original Stories and Sketches*. London, 1850.

Willis, Robert. *An Attempt to Analyse the Automaton Chess Player of Mr. De Kempelen*. London, 1821.

Windisch, Karl Gottlieb von. *Lettres sur le joueur d'echecs*. Basle, 1783 (translated as *Inanimate Reason*, London, 1784).

Zweig, Stefan. *The Royal Game*. London, 1944.

CHAPTER THREE

"Edison's Phonograph Dolls," *Scientific American*, April 26, 1890.

Barthes, Roland. *Camera Lucida*, tr. Richard Howard. London, 1993.

Bibliography

Bloch, Iwan. *The Sexual Life of Our Time*. London, 1923.

Bryan, George S. *Edison: The Man and His Work*. New York, 1926.

Clark, Ronald W. *Edison: The Man Who Made the Future*. London, 1977.

Coleman, Dorothy, Evelyn, and Elizabeth. *The Collector's Encyclopaedia of Dolls*. New York, 1968.

Dickson, William K. L. *The Life and Inventions of Thomas Edison*. London, 1894.

Dyer, Frank, and T. C. Martin. *Edison: His Life and Inventions*. New York, 1929.

Edison, Thomas Alva. "The Woman of the Future," *Good Housekeeping*, October 1912.

Fales-Northrop. "The Talking Doll," *Harper's Young People*, February 28, 1891.

Ford, Henry. *My Friend Mr. Edison*. London, 1930.

Formanek-Brunell, Miriam. *Made to Play House*. New Haven, Conn., 1993.

Frow, George. *Cylinder Phonograph Companion*. Woodland Hills, Calif., 1994.

Garbit, F. J. "Edison's Speaking Phonograph, or 'Talking Machine,'" in *Half-Hour Recreations in Popular Science*. Boston, 1879.

Haraway, Donna. "A Manifesto for Cyborgs," *Socialist Review* 15, no. 2 (March–April 1985).

• Harbou, Thea von. *Metropolis*. London, 1927.

Harper's Young People, January 27, 1891.

Hollingshead, John. *My Lifetime*, vol. 1. London, 1895.

Hopkins, Albert. *Magic: Stage Illusions and Scientific Diversions*. London, 1897.

Bibliography

Hounshell, David A. *From the American System to Mass Production, 1800–1932*. Baltimore, 1984.

Hughes, Ted. *Tales from Ovid*. London, 1997.

Israel, Paul. *Edison: A Life of Invention*. New York, 1998.

Jehl, Francis. *Menlo Park Reminiscences*. New York, 1990.

Jenkins, Rosenberg, et al. *The Papers of Thomas Alva Edison*. Baltimore, 1989.

Kempelen. *Le Mécanisme de la parole*. Vienna, 1791.

Lynd, William. *Edison and the Perfected Phono*. Tunbridge Wells, Eng., 1891.

McFadden, Sybill. "Mr. Edison's Astonishing Doll," *Hobbies*, August 1983.

Marx, Karl. *Selected Writings*, ed. David McLellan. Oxford, 1977.

Marx, Karl, and Frederick Engels. *Manifesto of the Communist Party*. London, 1888.

Michelson, Annette. "On the Eve of the Future: The Reasonable Facsimile and the Philosophical Toy," *October* 29 (Summer 1984).

Ovid. *Metamorphoses*. London, 1955.

Pacteau, Francette. *The Symptom of Beauty*. London, 1994.

Preece, Sir W. H. *The Phonograph*. London, 1878.

Raitt, A. W. *The Life of Villiers de l'Isle-Adam*. Oxford, 1981.

Rodgers, Daniel T. *The Work Ethic in Industrial America, 1850–1920*. Chicago, 1974.

Runes, Dagobert, ed. *The Diary and Sundry Observations of Thomas Alva Edison*. New York, 1948.

Schlereth, Thomas J. *Victorian America, 1876–1915: Transformations in Everyday Life*. New York, 1991.

Seltzer, Mark. *Bodies and Machines*. London, 1992.

Bibliography

Simonds, William A. *Edison: His Life, His Work, His Genius.* New York, 1940.

Simonelli, Yolanda. "Edison's Talking Doll—An Idea That Failed," *Doll Reader,* June–July 1988.

Tate, Alfred. *Edison's Open Door.* New York, 1938.

The Voice by Wire and Post-Card: All About the Telephone and Phonograph. London, 1878.

Thomas Alva Edison Papers, microfilm edition, part 3, "Edison Phonograph Toy Manufacturing Co.," 1889–1892.

Villiers de l'Isle-Adam. *Oeuvres Complètes,* vol. 1. Paris, 1986.

Wagner, A. F. *Recollections of Thomas Alva Edison.* London, 1991.

Warren, Waldo P. "Edison on Invention and Inventors," *Century Magazine* 82 (1911).

Welch, Walter L., and L. B. S. Burt. *From Tinfoil to Stereo.* Gainesville, Fla., 1994.

Wollstonecraft, Mary. *A Vindication of the Rights of Woman.* London, 1986.

CHAPTER FOUR

Altick, Richard. *The Shows of London.* Cambridge, Mass., 1978.

Arias, Pierre. "Méliès Mécanicien," in Madeleine Malthête-Méliès, ed., *Méliès et la naissance du spectacle cinématographique.* Paris, 1984.

Barnouw, Erik. "The Magician and the Movies," *American Film,* April 1978.

Barnum, Phineas Taylor. *Struggles and Triumphs, or Forty Years' Recollections.* London, 1869.

Bibliography

Baschet, Roger. *Le Monde Fantastique de Musée Grévin*. Paris, 1982.

Bergson, Henri. *Le Rire: essai sur la signification du comique*. Paris, 1912.

Bessy, Maurice, and Lo Duca. *Georges Méliès, Mage*. Paris, 1961.

Braun, Marta. *Picturing Time: The Work of Etienne-Jules Marey (1830–1904)*. Chicago, 1992.

Callas, P., and D. Watson, eds. *Phantasmagoria: Pre-Cinema to Virtuality*. Museum of Contemporary Art, Sydney, 1996.

Cherchi Usai, Paolo. *Burning Passions: An Introduction to the Study of Silent Cinema*. London, 1994.

———, ed. *A Trip to the Movies: Georges Méliès, Filmmaker and Magician (1861–1938)*. Rochester, N.Y., 1991.

Clément, Catherine. *Les Fils de Freud sont fatigués*. Paris, 1978.

Dagognet, François. *Etienne-Jules Marey: A Passion for the Trace*. New York, 1992.

Didi-Huberman, Georges. *Invention de l'hystérie: Charcot et l'iconographie photographique de la Salpêtrière*. Paris, 1982.

Dif, Max. *Histoire illustrée de la prestidigitation*. Paris, 1986.

Evans, Henry Ridgely. *Magic and Its Professors*. Philadelphia, 1902.

Evans, Martha Noel. *Fits and Starts: A Genealogy of Hysteria in Modern France*. Ithaca, N.Y., 1991.

Ezra, Elizabeth. *Georges Méliès: The Birth of the Auteur*. Manchester, Eng., 2000.

Fielding, Raymond, ed. *A Technological History of Motion Pictures and Television*. Berkeley, Calif., 1967.

Finger, Stanley. *Minds Behind the Brain: A History of the Pioneers and Their Discoveries*. New York, 2000.

Fischer, Lucy. "The Lady Vanishes: Women, Magic and the Movies," *Film Quarterly*, Fall 1979.

Francesco, Grete de. *The Power of the Charlatan*. New Haven, Conn., 1939.

Frazer, John. *Artificially Arranged Scenes*. Boston, 1979.

Freud, Sigmund. "The Uncanny," in *The Penguin Freud Library*, vol. 14. London, 1990.

————. "Charcot," in *Collected Papers*, vol. 1. London, 1924.

Goetz, C., M. Bonduelle, and T. Gelfand. *Charcot: Constructing Neurology*. New York, 1995.

Gross, Kenneth. *The Dream of the Moving Statue*. Ithaca, N.Y., 1992.

Hammond, Paul. *Marvellous Méliès*. London, 1974.

————. "Clinical Automatism" (unpublished manuscript).

Hendricks, Gordon. *The Edison Motion Picture Myth*. Berkeley, Calif., 1961.

Herbert, Stephen, and Luke McKernan, eds. *Who's Who of Victorian Cinema*. London, 1996.

Houdini, Harry (Erich Weiss). *The Unmasking of Robert-Houdin*. New York, 1908.

Kleist, Heinrich von. "On the Marionette Theatre," in Parry, Idris, ed., *Essays on Dolls*. London, 1994.

Lamb, Geoffrey. *Victorian Magic*. London, 1976.

Lefebvre, T., J. Malthête, and L. Mannoni, eds. *Lettres d'Etienne-Jules Marey à Georges Demenÿ, 1880–1894*. Paris, 1999.

Leuba, Marion, ed. *Marey: Pionnier de la synthèse du mouvement*. Beaune, 1995.

Lumière, Louis and Auguste. *Letters*. London, 1995.

Malthête, Jacques. *Méliès: Images et illusions*. Paris, 1996.

Malthête-Méliès, Madeleine. *Méliès l'enchanteur*. Paris, 1973.

Bibliography

Mannoni, Laurent. *Etienne-Jules Marey: La Mémoire de l'oeil*. Paris, 1999.

——. *Georges Demenÿ: Pionnier du cinéma*. Paris, 1997.

Marey, E.-J. *La Circulation du sang*. Paris, 1881.

——. *Le Vol des oiseaux*. Paris, 1890.

——. *Le Mouvement*. Paris, 1894.

Maskelyne and Cooke, The Royal Illusionists. London, 1875.

Maskelyne, Jasper. *White Magic: The Story of Maskelynes*. London, 1936.

Maskleyne, John Nevil. *Modern Spiritualism*. London, 1875.

——. *The Fraud of "Theosophy" Exposed*. New York, 1912.

Méliès, Georges. "Mes Mémoires," in Maurice Bessy and Lo Duca, *Georges Méliès, Mage*. Paris, 1961.

——. "Vues Cinématographiques," in *Exposition commemorative du centenaire de Georges Méliès*. Paris, 1961.

Merritt Crawford Papers, Film Studies Center, Museum of Modern Art, New York.

Musser, Charles. *Before the Nickelodeon*. Berkeley, Calif., 1991.

Muybridge, Eadweard. *Muybridge's Complete Human and Animal Locomotion*, 3 vols. New York, 1979.

Quévrain, Anne-Marie, and Marie-George Charconnet-Méliès. "Méliès et Freud: un avenir pour les marchands d'illusions?," in Madeleine Malthête-Méliès, ed., *Méliès et la naissance du spectacle cinématographique*. Paris, 1984.

Robert-Houdin, André Keime. *Robert-Houdin, le magicien de la science*. Paris, 1986.

Robert-Houdin, J.-E. *Memoirs of Robert-Houdin: King of the Conjurers*, trans. Lascelles Wraxall. New York, 1964.

Robertson, Gaspard. *Mémoires récréatifs*. Paris, 1831.

Bibliography

Robinson, David. *From Peep Show to Picture Palace: The Birth of American Film.* New York, 1996.

————. *Georges Méliès: Father of Film Fantasy.* London, 1993.

Sadoul, Georges. *Georges Méliès.* Paris, 1961.

————. *Histoire générale du cinema.* Paris, 1947.

Warner, Marina. "The Uses of Enchantment," in Duncan Petrie, ed., *Cinema and the Realms of Enchantment.* London, 1993.

————. "Through a Child's Eyes," in Duncan Petrie, ed., *Cinema and the Realms of Enchantment.* London, 1993.

————. "Women Against Women in the Old Wives' Tale," in Duncan Petrie, ed., *Cinema and the Realms of Enchantment.* London, 1993.

CHAPTER FIVE

"Daisy Doll, 3 foot 6, Admits She's Looking for a Husband," *New York Post,* April 12, 1937.

"Daisy, A Doll-Sized Ballerina," *Sarasota Herald Tribune,* July 4, 1965.

"Doll Family Adds Home Cut to Scale," *Sarasota Herald Tribune,* 1936.

"Harry Earles," obituary, *Variety,* December 25, 1985.

"Kurt 'Harry Doll' Schneider," obituary, *Sarasota Herald Tribune,* May 2, 1985.

"The Big Show," *Banner Line,* March 1, 1953.

Babbage, Charles. *Passages from the Life of a Philosopher.* London, 1864.

Baudelaire, Charles. "The Philosophy of Toys," in *Essays on Dolls.* London, 1994.

Bibliography

Baudrillard, Jean. *Simulations.* Paris, 1983.

Boruwlaski, Joseph. *Life and Love Letters of a Dwarf.* London, 1902.

————. *Memoirs of the Celebrated Dwarf.* London, 1788.

Brosnan, John. *The Horror People.* London, 1976.

Bryan, J., III. "Minute Man," *Collier's,* May 2, 1942.

Carroll, Lewis. *Alice's Adventures in Wonderland and Through the Looking-Glass.* Oxford, 1971.

Chindahl, George L. *A History of the Circus in America.* Caldwell, Idaho, 1959.

Cox, Stephen. *The Munchkins Remember: The Wizard of Oz and Beyond.* New York, 1989.

De Groft, A., M. Ormond, and G. Ray, eds. *John Ringling: Dreamer, Builder, Collector.* Sarasota, Fla., 1996.

De la Mare, Walter. *Memoirs of a Midget.* New York, 1922.

Dickens, Charles. *Sketches by Boz.* London, 1994.

Drimmer, Frederick. *Very Special People.* New York, 1973.

Evans, Walker. *Florida.* Los Angeles, 2000.

Fellner, Chris. "Don't Cry Hans!: The Life and Career of Harry Doll, Star of 'Freaks,' " *Freaks!,* May 1997.

Fiedler, Leslie. *Freaks: Myths and Images of the Secret Self.* New York, 1978.

Findlay, James A., and Margaret Bing. "Touring Florida Through the Federal Writers' Project," *Journal of Decorative and Propaganda Arts,* Florida theme issue, no. 23, 1998.

Florida: A Guide to the Southernmost State. New York, 1939.

Fournier, Edouard. *Histoire des jouets et des jeux d'enfants.* Paris, 1889.

Fricke, John, Jay Scarfone, and William Stillman. *The Wizard*

Bibliography

of Oz: The Official 50th Anniversary Pictorial History. New York, 1989.

Gould and Pyle. *Anomalies and Curiosities of Medicine*. New York, 1896.

Grismer, Karl H. *The Story of Sarasota*. Sarasota, Fla., 1946.

Harmetz, Aljean. *The Making of the Wizard of Oz*. New York, 1989.

Hilliar, W. J. "A History of the Circus Sideshow," in *Billboard*, March 23, 1918.

Johnston, Alva. "Profiles," *The New Yorker*, May 6, 1933.

————. "Sideshow People," *The New Yorker*, April 14, April 21, and April 28, 1934.

Kasson, John. *Amusing the Million: Coney Island at the Turn of the Century*. New York, 1978.

Lahr, John. *Notes on a Cowardly Lion*. New York, 1969.

LaHurd, Jeff. *Quintessential Sarasota: Stories and Pictures from the 1920s–1950s*. Sarasota, Fla., 1990.

Lentz, John. "The Revolt of the Freaks," in *The Bandwagon*, September–October 1977.

McCullough, Edo. *Good Old Coney Island*. New York, 1957.

McDonough, Michael. "Selling Sarasota: Architecture and Propaganda in a 1920s Boom Town," *Journal of Decorative and Propaganda Arts*, Florida theme issue, no. 23, 1998.

Murray, Marian. *Circus! From Rome to Ringling*. New York, 1956.

O'Brien, Esse Forrester. *Circus: Cinders to Sawdust*. San Antonio, 1959.

Paré, Ambroise. *Animaux, monstres et prodiges*. Paris, 1954.

Pilat, Oliver, and Jo Ranson. *Sodom by the Sea: An Affectionate History of Coney Island*. New York, 1941.

Bibliography

Plowden, Gene. *Those Amazing Ringlings and Their Circus.* Caldwell, Idaho, 1967.

Rilke, Rainer Maria. "Dolls: On the Wax Dolls of Lotte Pritzel," in *Essays on Dolls.* New York, 1994.

Robbins, Tod. *The Unholy Three.* New York, 1917.

———. "Spurs," *Munsey's Magazine,* 1923.

Roth, Hy, and Robert Cromie. *The Little People.* New York, 1980.

Rushdie, Salman. *The Wizard of Oz.* London, 1992.

Savada, Elias. "The Making of *Freaks*," *Photon,* no. 23, 1973.

Skal, David J., and Elias Savada. *Dark Carnival: The Secret World of Tod Browning—Hollywood's Master of the Macabre.* New York, 1995.

Stewart, Susan. *On Longing: Narratives of the Miniature, the Gigantic, the Souvenir, the Collection.* Durham, N.C., 1993.

Stowe, Harriet Beecher. *Palmetto Leaves.* Boston, 1873.

Swift, Jonathan. *Gulliver's Travels.* New York, 1967.

The Ringling Brothers and Barnum & Bailey Combined Circus Magazine and Daily Review, 1927 and 1933.

Thompson, C. J. S. *The History and Lore of Monsters.* London, 1996.

Thompson, Frederic. "Amusing the Million," in *Everybody's Magazine.*

Tucker, Albert. "The Strangest People on Earth," *Sarasota Sentinel,* July 7 and July 14, 1973.

Weeks, David C. *Ringling: The Florida Years, 1911–1936.* Gainesville, Fla., 1993.

White, Mitch. "Those Happy Little People," *The Bandwagon,* November 1957.

Wood, Edward. *Giants and Dwarfs.* London, 1868.

Index

Adelaide Gallery (London), 76, 216
Aeschylus, xv
Age of Reason, 31, 82
Agrippa, Cornelius, xvi
AI (film), xxvi
Albertus Magnus, xvi, 179
Alembert, Jean le Rond d', 21, 55
Alexandre, Aaron, 85–7, 91
Allen, George, 84–6, 89–92, 95–6, 98, 99
Allgaier, Johann Baptist, 86, 92
Altick, Richard, 82
American system, 116–17
anatomical models, 46–8
Anderson, Professor, 89, 90, 92, 98
androids, xiv–xv, xviii, 16–17, 19, 127; cinema and, 175–6, 191; in magic shows, 165–6, 179–81; *see also* robots; *specific automata*
Animatograph, 183
Antoine, Michel, 45
Aquinas, Thomas, xvi
Aragon, Louis, 209
Archytas of Taretum, xv
Arnould, Monsieur, 175, 181
artificial intelligence (AI), xx–xxvi
Artificially Arranged Scenes (John Frazer), 193
Athol, Duke of, 70
"Automata, The" (Hoffmann), 33, 63
Avril, Jane, 199

Babbage, Charles, 76, 216–17, 219
Baclanova, Olga, 235

Index

Bacon, Roger, 179

Baltimore Gazette, 75

Bancroft, Anne, 249

Barnum, P. T., 75–6, 81, 165, 181, 217, 240

Barnum & Bailey Circus, 248

Batchelor, Charles, 148

Baudelaire, Charles, 212, 215

Beethoven, Ludwig van, 73

Beireis, Gottfried Christoph, 33–7, 263

Bell and Tainter, 148

Bellmer, Hans, 152

Benjamin, Walter, 99, 109

Berg, Duke of, 36

Bergson, Henri, 191, 199

Bernhardt, Sarah, 199

Bertin, Nicolas, 20, 56–8

Beugnot, Count, 36–7

Bihéron, Marie Catherine, 47–8

Binet, Alfred, 104–5, 175

Birth of a Nation (film), 205–6

Blade Runner (film), xviii, xxiv, xxvi, 266

blindfold chess, 104–5, 107, 108, 175

Bloch, Iwan, 138–9

blood, circulation of, 13, 172; mechanical models of, 46–60, 113, 179

Boerhaave, Herman, 11

Boncourt (chess player), 87, 92

Bontems, Blaise, 39

Boruwlaski, Joseph, 220–1

Brabeck, Count, 36–7

Breazeal, Cynthia, xxii, xxiii, xxviii, 264

Brecht, Bertolt, 250–1, 255

Breton, André, 209

Brewster, David, 111, 125

Briggs, Lowell, 148

British Psycho-Analytical Society, 105

Brooks, Rodney, xxi, xxii, xxv–xxvi

Browning, Tod, 228, 234–8

Brunswick, Duke of, 33, 34

Buatier de Kolta, Joseph, 181, 192

Bungy, Thomas, 179

Buñuel, Luis, 209

Café de la Régence (Paris), 84, 85, 87, 91

Cagliostro, Count Alessandro, 167

Calmels, Eugène, 181

Capek, Karel, xviii, xix, xxiii
Catherine the Great, Empress of Russia, 48, 62
Centre National de la Cinématographie (CNC), 189
Chaney, Lon, 228–30, 233, 260
Chaplin, Charlie, 118
Chapuis, Alfred, 38
Charcot, Jean-Martin, 197–201
Chardin, Jean-Baptiste-Siméon, xx
Chatelet Theatre (Paris), 186
Chess Player, Automaton, xxiv, 60–109, 112, 124, 125, 165, 167, 215, 217, 240; attempts to discredit, 67–72, 74–5, 171; creation of, 60–2, 64; operators of, 84–102, 107–9; Poe's article on, 76–9, 213; revelation of secret of, 81–3, 100–1
Chien Andalou, Un (film), 209
child automata, xiii–xv, xx, 3–6, 215–17
Chinese Museum (Philadelphia), 81, 96
Chovet, Abraham, 50, 52

Christina, Queen of Sweden, 3
Cideville, Pierre-Robert Le Cornier de, 53–4
cinema, 167–76, 183–212, 219; Edison and, 114, 130, 131, 144, 168, 170–1, 174, 182, 183, 204–5; magic and, 183–5, 187, 190–7, 201; midgets in, 228, 242–4; origins of, 168–75; sound, 223–4; see also titles of specific films
Cinémathèque Française, 174, 189
Cinematograph, 175, 182–3
Circulation du Sang, La (Marey), 172
circulatory system, see blood, circulation of
circus performers, 214, 216, 219, 224–7, 235, 237, 243–6, 248, 251
Clarke, Arthur C., 164
Clément, Catherine, 199
Cody, "Buffalo Bill," 222
Cog, xxi–xxiii
Cohen, John, xvii, 92
Collier's magazine, 220, 226, 245, 246
Collinson, Thomas, 70–1, 125

Index

Communist Manifesto (Marx and Engels), 117

Condillac, Etienne Bonnot de, 16, 265, 266

Condorcet, Marquis de, 41, 58

conjuring, *see* magic

Conservatoire National des Arts et Métiers (Paris), 39

Cook, Captain James, 164–5

Cooke, George Alfred, 165, 166

Coysevox, Antoine, 21, 22, 45

Crachami, Caroline, 217, 221

Crawford, Joan, 261

Crawford, Merritt, 207, 211–12

Curtius, Philippe, 46

D'Alcy, Jehanne, *see* Manieux, Fanny

Dalí, Salvador, 209

Darwin, Charles, 165

Darwin, Erasmus, 132

Davenport Brothers, 166, 182

Decremps, Henri, 70, 82, 166

Deep Blue, xxiv, 109

Defence, The (Nabokov), 103–4

Delannoy, Henri Auguste, 100

Demenÿ, Georges, 175

Descartes, René, 3–14, 24, 202, 215, 266

Deschapelles, Louis Honoré Lebreton, 85, 100

Desfontaines, Abbé, 27–8

Deslandes, André-François, 16

Desnoues, Guillaume, 47

Devant, David, 183

Dick, A. B., 149–53, 155–9, 161

Dickens, Charles, 127, 221, 239

Dickson, W. K. L., 119, 121, 124, 129, 168, 171

Diderot, Denis, 11, 16, 21, 219

Dietz, Georges, 37–8

Disney, Walt, 209

DNA, xxvi

Doll Family, 214–16, 218–19, 223–38, 241, 243–7, 251–63

dolls, xv, xxv–xxvi, 214–15, 217–18; talking, xxvi, 118–24, 127, 128, 144–5, 148–63, 185, 196

Doyon, André, 17, 40, 42

Dracula (film), 228

Drawing Lesson, The (film), 196

du Barry, Madame, 25

Duchamp, Marcel, 96, 97

duck automaton, xviii, 15, 17, 26–30, 32–33, 35–40, 46, 112, 123, 178, 180, 219

Dumoulin, 32

dwarfs, 217, 220–1, 239

Dyer, Frank, 115

Earle, Jack, 226

Earles family, 221–4; *see also* Doll Family

Eastman, George, 171

Ebert, Franz, 228

Eck, Johnny, 235

Eco, Umberto, xvii

Edison, Marion, 148

Edison, Mina, 146–8

Edison, Thomas, 111–18, 145–48, 176, 215, 219, 263; moving pictures and, 114, 130, 131, 144, 168, 170–1, 174, 182, 183, 204–5; novel about, 131–8, 140–5, 151, 152; phonograph of, 39–40, 118, 127–31, 144, 167, 170, 218; Ringling and, 248–9; talking doll of, xxvi, 118–24, 127, 128, 144–5, 148–63, 185, 196

Egyptian Hall (London), 164–5, 167–8, 176, 181–3, 235, 263

Eiffel, Gustave, 114

Ellis, Cab, 120

Eluard, Paul, 209

Encyclopédie, 11, 21, 30, 42

Engineering magazine, 129, 130

Euphonia, 127–8

Enlightenment, xvii, xix, 7, 12, 17, 22, 41, 42, 59, 167, 168

Evans, Walker, 225

Eve of the Future, The (Villiers de l'Isle-Adam), xix, 131–8, 140–5, 151, 152, 264

Everybody's magazine, 222

Extraordinary Illusions (film), 196

Faber, Professor, 127–8

Fagon, Guy-Crescent, 49

Fantasmagoria, 166

Faust (Goethe), xvi, 132

Fiedler, Leslie, 239

Figaro, 144

Firestone, Harvey, 113

Fischer, Lucy, 192, 194

Fitzgerald, F. Scott, 237

Fleury, Cardinal de, 44

Flute Player, Automaton, 15, 17, 21–7, 32, 35, 60, 127, 240, 264, 265

Fontana, Felice, 47

Ford, Harrison, 266

Ford, Henry, 113, 162

Fordian mass production, 116–17

Formanek-Brunell, Miriam, 122–3, 149
Fournier, Edouard, 218
Four Troublesome Heads (film), 202
Francesco, Grete de, 167
Francini, Tomaso, 14
Francis, Mrs. H. M., 160–1
Frankenstein (Shelley), xv, 132, 168, 197
Franklin, Benjamin, 84, 125
Freaks (film), 228, 234–8, 251, 253, 261, 262
Frederick the Great, King of Prussia, 12, 30–2
French Revolution, 227
Fresneau, François, 56–8
Freud, Sigmund, xiv–xv, 138, 139, 191, 198–201, 224–5

Gable, Clark, 261
Gainsborough, Thomas, 67
Galerie Vivienne (Paris), 176
Galileo, 9
Garbo, Greta, 243, 261
Garland, Judy, 241–2
Gaumont film company, 205
Gélis, Edouard, 38
George III, King of England, xiv, 73

Gilson, Paul, 210
Godwin, William, xv
Goethe, Johann Wolfgang von, xvi, 34–5, 40, 132
Good Housekeeping, 147
Gravelot, Hubert, 22
Great Expectations (film), 249
Greek mythology, xv, xix, 16, 17
Griffith, D. W., 205–6, 209
Grunewald, Marlene Townsend, 254–6, 260

Haller, Albrecht, 13
Hardy, Oliver, 228, 229
Harper's Young People, 119, 122, 161
Harvey, William, 13, 49
Hasbro Toys, xxv–xxvi
Hawthorne, Nathaniel, 130
Hero of Alexandria, xvi
Herrmann, Leon, 235
Heth, Joice, 75, 240
Hildesheim, Bishop of, 34
Hilton sisters, 235, 238
Hoffmann, E. T. A., xix, 33, 63, 142, 182, 191, 265, 266
Hollingshead, John, 127–8
Homme machine, L' (La Mettrie), 11–15

Index

Hopkins, Albert, 120–2, 184–5
Hôtel de l'Echiquier (Paris), 85–7
Houdini, Harry, 38
Household Words, 127
Hughes, Ted, 138
Hutton, Thomas, 71
Huyghens, Christian, 14, 15
Huyssen, Andreas, xix
Hyams, Leila, 238, 261
hysteria, 197–201

Illustrated London News, 169
Industrial Revolution, 42
Inquisition, 9
Instruction in Natural Magic (Wiegleb), 33, 63
Insull, Samuel, 153
International Congress of Film Editors, 205
Interpretation of Dreams, The (Freud), 201
iRobot, xxv
Israel, Paul, 112, 144

Jacquard loom, 41
Jacques, William, 148
Jacquet, Gaspard, 43
Janet, Pierre, 201
Jans, Hélène, 5

Jaquet-Droz, Henri-Louis, xiv, 24
Jaquet-Droz, Pierre, xiv, xv, xvii, xx, 8
Jerdan, William, 217
Johnston, Alva, 216, 226
Jones, Ernest, 105–7
Joseph II, Emperor of Austria, 61
Joseph-Josephine, 235
Journal des Savants, 50
Jumeau, Emile, 150–2, 157, 218
Jussieu brothers, 55

Kasparov, Gary, xxiv, 109
Kato, Ichiro, 264
Kelley, Charles, 242
Kelly, J. S., 153–4
Kempelen, Wolfgang von, xxvi, 60, 76, 81, 83, 176, 216, 219; Chess Player of, xxvi, 60–72, 75, 78, 79, 84, 86, 94, 101, 110, 112, 124, 165, 167, 178, 217; speaking machine of, 124–7, 134, 265
Kinetograph, 130, 182, 204
Kinetoscope, 130, 171, 175, 182
Kismet, xxi–xxii, xxv
Kleist, Heinrich von, 191

Index

Kodak camera, 171
Kookoo the Bird Girl, 235
Korsten, Lucien, 183
Krebs, Nita, 256–7

La Condamine, Charles Marie
    de, 55–8
Lahr, Bert, 242
La Mettrie, Julien Offroy de,
    10–17, 19, 30, 31, 166, 172,
    173, 200
Lancaster, Burt, 261
Lang, Fritz, 118
La Peyronie, François de, 45, 49
Lathrop, George, 130–1
Laurel, Stan, 229
Le Cat, Claude-Nicolas, 20, 50,
    52–4, 58
Lenya, Lotte, 250–1
Leonardo da Vinci, 47
LeRoy, Mervyn, 242
Letters on Natural Magic (Brew-
    ster), 111
Lewis, William, 84–8, 97, 100,
    107–8
Liaigre, Lucien, 17, 40, 42
Library Company of Philadel-
    phia, 83–4, 96
Lilliputian Minuet, The (film),
    196

Linnaeus, Carolus, 55
Literary Gazette, 217
Lives of the Necromancers
    (Godwin), xv
Living Playing Cards, The
    (film), 196
Lloyd, Harold, 259, 261
Londe, Albert, 198–9
London, Museum of, 165
Louis XIV, King of France, 30,
    43, 49
Louis XV, King of France, xiv,
    25, 40, 43–6, 50, 55, 56, 58,
    181, 263
Louis-Philippe, King of France,
    100, 181
Lugosi, Bela, 228
Lumière brothers, 175, 182–4,
    189
Luna Park (Coney Island),
    222–3

Macquer, Pierre-Joseph, 58
Maelzel, Johann Nepomuk,
    72–3, 75–8, 80, 83–4,
    87–92, 95–8, 108, 124, 193,
    216, 218
Magazine pittoresque, 81, 83, 100
magic, 34, 38, 62–3, 165–7,
    175–82, 199, 207, 208, 235;

cinema and, 183–5, 187, 190–7, 201; use of androids in, 165–6, 179–81

Magic (Hopkins), 184

Magic Lantern, The (film), 196

Malthête-Méliès, Madeleine, 207, 210, 211

Manieux, Fanny (Jehanne D'Alcy), 182, 192, 207–8, 211

Man with the Rubber Head, The (film), 202

Manzolini, Anna and Giovanni, 47

Mareschal, Georges, 45, 49

Marey, Etienne-Jules, 170–5, 184, 195, 198, 203

Maria Theresa, Empress of Austria, 61

Martin, T. C., 115

Marx, Karl, 117

Maskelyne, John Nevil, 165–7, 183, 206, 220

Massachusetts Institute of Technology (MIT), xx–xxiii, xxv, xxvi, 264

mass production, 116–17, 219

Mauclaire, Jean, 209

Maupassant, Guy de, 199

McGovern, E., 162

McLaglen, Victor, 229, 233

Méliès, Georges, xxvi, 167, 175–7, 181–97, 199–212, 215, 219, 222; family of, 204–7, 211

Metro-Goldwyn-Mayer (MGM), 229, 237, 238, 242, 243

Metropolis (film), 118, 122

Michelangelo, 47

Microscopic Dancer, The (film), 196

midgets, 214, 217, 220, 222, 227–8, 239; circus, 227; in films, 228, 241–3, 256; *see also* Doll Family

mimeograph, 150

Mitchell, John Kearsley, 80

Modern Times (film), 118

Morphy, Paul, 105–7

Motion Picture Patents Company (MPPC), 204–5

Mouret, Jacques-François, 88, 91, 97, 100–2, 107, 109

moving pictures, *see* cinema

Munsey's Magazine, 234

Musée des Arts et Métiers (Paris), 206

Index

Musée Grévin (Paris), 176–7

Museum of Modern Art (New York), 208

Musical Lady, xiv, 24

Muybridge, Eadweard, 168–70, 173

Mysterious Dislocations (film), 200–1

Mysterious Portrait, The (film), 201

Nabokov, Vladimir, 102–4

Napoleon, Emperor of France, 25, 36, 62, 86, 165

National Gazette, 83

National Hotel (New York), 88–9

Natural History of the Soul, The (La Mettrie), 11

Neuromancer (film), 265

New Extravagant Battles (film), 201

New York Daily Graphic, 112

New Yorker, 216, 226, 227

New York Post, 89, 129, 244–5

New York Times, 129, 233

New York World, 144

Nicolai, Christian Friedrich, 32–3, 38

Novarro, Ramon, 261

One-Man Band, The (film), 201

Orange Chronicle, 120

Ott, Fred, 171

Ovid, 138, 190

Palamède, Le (chess periodical), 83, 101

Panharmonicon, 73

Paracelsus, xvi

Paris Expositions, 38, 114, 118, 131, 144, 156, 170, 180–1

Paris Opera, 186

Partridge, Horace, & Company, 160

Pathé, Charles, 205–7

Pathfinder mission, xxi

Paul, Robert William, 183

Peale, Charles Wilson, 81, 263

Peter the Great, Tsar of Russia, 241

Philadelphia Gazette, 80

Philidor, François André Danican, 62, 84, 100, 105

Philosophical and Mathematical Dictionary (Hutton), 71

Philosophical Investigations (Wittgenstein), 60

phonograph, 39–40, 118, 127–31, 144, 167–8, 170, 218

photography, 168; sequential,
168–70, 173–4 (*see also*
cinema); trick, 184–5
Poe, Edgar Allan, 76–9, 213
Popular Science Monthly, 127,
128
Potter, Henry C., 118
Praxinoscope, 177
Prelude, The (Wordsworth),
213
Pulitzer, Joseph, 162
Pygmalion and Galatea (film),
187–90, 194

Quesnay, François, 49–52

Racknitz, Joseph Friedrich
Freiherr zu, 60, 70–2, 78,
81, 94
Rameau, Jean-Philippe, 16
Raynaly, Monsieur, 175, 181
Récamier, Madame, 146
Reichsteiner, Johann-
Bartholomé, 37–8
Reulos, Lucien, 183
Reynaud, Emile, 177
Richebourg (court midget),
227–8
Rilke, Rainer Maria, 240
Ringling, John, 243, 247–8, 250

Ringling Brothers Circus, 216,
224–5, 235, 245, 252
Roach, Hal, 228
Robbins, Tod, 228–9, 231–2, 235
Robert-Houdin, Jean-Eugène,
38, 77, 177–84, 204, 206,
208, 220
Robert-Houdin Theatre (Paris),
175, 177, 181–3, 187, 206,
207
Robertson, Gaspard, 166
Robin, Henri, 165
Robinson, David, 184
Robocop (film), xxvi
robots, xvii–xix, xxi–xxiii, 264–5
Rossum's Universal Robots
(Capek), xviii, xxiii
Rousseau, Jean-Jacques, xx, 16
Royal Game, The (Zweig),
102–4
rubber, 55–8, 113, 249
Runyon, Louis, 245
Ruskin, John, 118

Salle Playel (Paris), 209–11
Salpêtrière (Paris), 197–200
"Sandman, The" (Hoffmann),
xix, 33, 63, 136, 142, 182,
191
Sarasota Herald Tribune, 246–7

Schaffer, Simon, 17, 109

Schlumberger, Wilhelm, 90–2, 96, 97

Schmidt, Charles F., 89–91, 95, 98–9

Schneider family, 214–18, 220–4, 240, 262; *see also* Doll Family

Scientific American, 120, 185

Sennett, Mack, 229

Shelley, Mary, xv, 132

Shore, Dinah, 261

silk industry, mechanization of, 19–20, 40–3, 51, 109, 116

Simon & Halbig, 157

Singer, Leo, 228, 241–3

Smith, Lloyd P., 96

Sneeze, The (film), 171

Southern Literary Messenger, 76

speaking machines, 124–8, 134, 265; *see also* dolls, talking; phonograph

Stanford, Leland, 168–9

Star Films, 204–5

Steiner, Jules Nicolas, 218

Stewart, Susan, 240

Stowe, Harriet Beecher, 250

Studies on Hysteria (Freud and Breuer), 201

Studio 28 (Paris), 209

Takanishi, Atsuo, 264, 265

Takanobu, Hideaki, 265

Tate, Alfred, 112, 114, 150, 156, 159, 161

Ten Ladies in One Umbrella (film), 196

Terminator (film), xxvi

Thicknesse, Philip, 67–70, 171

Thompson, Frederic, 222–4

Thuillier, Madame, 209

Thumb, Tom, 165, 181, 217, 240

Torrini, 177–8

Toulouse-Lautrec, Henri de, 199

Toys and Novelties, 122, 123

Treatise on Man (Descartes), 7, 9, 10

Trembley, Abraham, 13

Trip to the Moon, A (film), 190

Turing test, xxiii–xxv

Turk, *see* Chess Player

Tussaud, Madame, 25, 46, 176

2001 (film), xxvi

Uncanny, the, xiv–xv, 168, 182, 199–201, 214–15, 217, 220

Unholy Three, The (film), 228–34, 238, 251, 260

Up-to-Date Conjuror, An (film), 194–5

Index

Vanishing Lady, The (film), 191–2, 194

Variety, 246

Vartanian, Aram, 15–16, 31

Vaucanson, Jacques de, xxvi, 14, 17–43, 45, 82, 113, 139, 156, 172, 176, 179–80, 206, 215, 218, 219, 266; blood-circulating machine of, 46–60, 113, 179; duck of, xviii, 15, 17, 26–30, 32–33, 35–40, 46, 112, 123, 178, 180; Flute Player of, 15, 17, 21–7, 32, 35, 60, 127, 240, 264, 265; and mechanization of silk industry, 19–20, 40–3, 51, 109, 116

Verne, Jules, 222

Vesalius, Andreas, 134

Victor, Henry, 235

Villard, Henry, 162

Villiers de l'Isle-Adam, xix, 131–8, 140–5, 147, 151, 152, 154, 264, 266

Vindication of the Rights of Women, A (Wollstonecraft), 111

Vitascope, 114

Voisin, Emile, 176

Voltaire, 16, 17, 20, 27, 29, 30, 51, 53

Voyage Through the Impossible, The (film), 190

Wabian, 265

Walker, George, 96, 100

Waseda University, xxviii, 264

Washington, George, 75

waxworks, 25, 39, 46–7, 176–7

Weill, Kurt, 250–1

Westworld (film), xxvi

Weyle (chess player), 87, 91

Wiegleb, Johann Christian, 33, 63

Williams, Peter Unger, 108

Willis, Robert, 73–5, 77

Wilmot, Eddie, 227

Windisch, Karl Gottlieb von, 64–8, 72, 78–9, 124–6

Wittgenstein, Ludwig, 60

Wizard of Oz, The (film), 241–3, 245, 252, 255, 256

Wollstonecraft, Mary, 111

women: as Charcot's patients, 197–201; Edison's attitude toward, 145–8; in Méliès's films, 191–7; perfect, creation of, xix, 131–45

Wood, Edward, 221

Index

Wordsworth, William, 213
Works Project Administration, 250
World War I, 206, 218
World War II, 102

Zeuxis, 148–9, 151
Ziegfeld, Florenz, 241
Zoopraxiscope, 169–70
Zumbo, Gaetano, 47
Zweig, Stefan, 102–4

A NOTE ABOUT THE AUTHOR

Gaby Wood was born in 1971 and attended Cambridge University. She has been a regular contributor to *The Guardian* and the *London Review of Books*. She is the author of a short work of nonfiction, *The Smallest of All Persons Mentioned in the Records of Littleness,* and is now living in London, where she is a staff writer for *The Observer.*

A NOTE ON THE TYPE

Pierre Simon Fournier *le jeune,* who designed the type used in this book, was both an originator and collector of types. His services to the art of printing were his design of letters, his creation of ornaments and initials, and his standardization of type sizes. His types are old style in character and sharply cut. In 1764 and 1766 he published his *Manuel typographique,* a treatise on the history of French types and printing, on typefounding in all its details, and on what many consider his most important contribution to typography—the measurement of type by the point system.

Composed by
North Market Street Graphics
Lancaster, Pennsylvania

Printed and bound by
Berryville Graphics
Berryville, Virginia

Designed by
Soonyoung Kwon